Conned Again, Watson!

also by Colin Bruce in Perseus Publishing:

THE EINSTEIN PARADOX, AND OTHER SCIENCE
MYSTERIES SOLVED BY SHERLOCK HOLMES

Conned Again, Watson!

Cautionary Tales of Logic, Math, and Probability

Colin Bruce

PERSEUS PUBLISHING

Cambridge, Massachusetts

Library of Congress Cataloging-in-Publication Data is available

ISBN 0-7382-0345-9

Perseus Publishing is a member of the Perseus Books Group.

Find us on the World Wide Web at http://www.perseuspublishing.com

Perseus Publishing books are available at special discounts for bulk purchases
in the U.S. by corporations, institutions, and other organizations. For more
information, please contact the Special Markets Department at HarperCollins
Publishers, 10 East 53rd Street, New York, NY 10022, or call 1-212-207-7528.

Set in 11-point Garamond Light by Perseus Publishing Services

First printing, November 2000

1 2 3 4 5 6 7 8 9 10—03 02 01

Contents

Preface

WE ALL LOSE TIME AND money every day to bad decisions. Often, we are not even aware of it. We continue in blissful ignorance, happy in the illusion that our native common sense is doing a good job of guiding us.

I was painfully reminded of this a few months ago. My friend Jo Keefe, a graduate student in decision theory at the London School of Economics, rang one afternoon.

"Colin," she said in dulcet tones, "I've got some questions here that should be fun for you. Professor X assigned a class exercise. We've each got to telephone half a dozen people we know and ask them what they think the chance of winning some simple bets would be. You're not supposed to work it out mathematically; just make a guess on the spot. Can I make you one of my panel?"

"Go ahead," I said confidently, thinking privately: "Ha! Think they can fool *me*, do they?"

The questions sounded reasonable enough, and I didn't hesitate to give rough answers. A week later, I discovered that one my guesses had been out by a factor of 10. It was little consolation to hear that Jo's other contacts, mostly Oxford mathematics graduates like herself, had been wrong by even larger multiples.

Now, of course Professor X is a fiend in human form, but the exercise was a timely reminder that we can all trip up on apparently simple choices, especially if probability or statistics is involved. And I have friends, as you will, who have lost more than pride through similar mistakes: commerce is run by people just as cunning as Professor X but with fewer scruples. I decided to do something about it.

There are math books and business books that between them cover the topics described here, but for many people they are rather dry reading. I have always enjoyed the ancient tradition of teaching with stories that contain a dire warning, like Aesop's Fables. Tell a tale of someone whose mistake has really awful consequences, and the moral becomes easy to remember.

Here, then, is a set of modern cautionary tales in the form of Sherlock Holmes stories. Please read them just for fun. If in the process you soak up a little knowledge about how to take life's gambles, and how not to be deceived by statistics and the modern-day con games all about us, I will be delighted.

Acknowledgments

MY GREATEST THANKS GO TO Claire Newman, who read the book as it progressed and made countless insightful suggestions for improvement. My editors Amanda Cook and Connie Day gave valuable advice. I am also grateful to Jo Keefe, to my sister Belinda, and to members of the Oxford University student mathematical society Invariant for many fruitful discussions.

Conned Again, Watson!

1

The Case of the Unfortunate Businessman

Iᴛ ᴡᴀs ᴛʜᴇ ᴍᴀɴ's ɢʜᴀsᴛʟʏ expression that first drew my attention to him. Eyes stared out from a lined and haggard face as he strode rapidly through the bustling crowds of Oxford Street, causing people laden with shopping to dodge and snarl at him. A moment later came the shock of recognition. I placed myself in his path and raised my arms to halt him.

"Cousin James," I cried. "How are you, man? Upon my word, you almost ran me down there!"

He looked at me, but my forced cheeriness brought no response from him. He broke his stride only for a moment, then headed on. I had almost to run to keep up with him.

He glanced sideways. "Get away from me, John. As to how I am, I am a ruined man. But worse, I am bringing others down with me. Poor McFarlane—he has had it as well, and all my fault. Go away, John, I am like a plague carrier. You look to be doing fine, and I will not drag family into this."

He turned his head away. I was totally taken aback. The last time I had seen him, just two months before, he had admittedly been looking somewhat careworn; but it was hardly surprising, for the occasion had been his father's funeral. In reality he had not got on well with his father, a self-made businessman who had shown a barely veiled contempt for his expensively educated eldest son. Certainly, nothing would have led me to expect this despair!

I grasped his arm. "You must think very little of me, James, if you think I would leave you in this state. Come, let us get some brandy down you."

I saw we were coming up to the Three Horseshoes pub. It is not the kind of place I would be seen in ordinarily, but this was no time to be particular. I steered my cousin across the sawdust-covered floor to the bar and ordered a double brandy. I watched a little color return to James's cheeks as he sipped it.

"Look, I promise I will not interfere except insofar as you permit it. But let me tell you as a doctor, the old saw 'a trouble shared is a trouble halved' is very true. Half the patients who consult me are helped more by a sympathetic ear than by any medicine I can prescribe for them. Now, whatever has brought you to this pass?"

He hesitated. "I don't deny it would be a relief to unburden myself, John. But I must have your promise, not only that you will not interfere, but that you will not breathe a word of what I have to say to another living soul. I have been unwise, John, incredibly unwise."

"You have my word on it."

He led me to a small corner table and glanced around: we were obviously quite safe from being overheard. He took a further swig of the brandy, and began: "You know, of course, that upon my father's death I inherited Watson's Cabs."

I nodded. Splendid in their livery of green with a large gold "W" on the side, his father's hansom cabs had been a familiar feature of the London landscape for as long as I could remember, a reminder that at least one of our family had been able to achieve financial success.

"Well, it was never my ambition to be a businessman, but I accepted the hand life dealt me and resolved to do my best. As well as supervising things day-to-day, I purchased some of the latest American books on business management and resolved to apply the principles they described. It is remarkable, John, how the Americans have reduced company management to an exact science. It became apparent to me that my father had been running things in a quite old-fashioned way. I envisaged doubling the profits of our traditional family business!"

He shook his head sadly. "But it seems I am not the man my father was. Far from increasing profits, my decisions seemed to diminish, then eliminate, them. A month ago, my accountant started to talk in terms of liquidation. I was desperate. How would I provide for my family if that happened? And then McFarlane came up with his scheme.

"He was an old friend of my father's. Actually, I met him for the first time at the funeral, but he told me they had been close from boyhood. He dropped in at the business from time to time, offering a sympathetic ear and fatherly advice, and refusing to accept more than a cup of coffee in return. I came to trust him, and in fact he was the only living soul who knew just how badly things were going.

"He told me he wished he could help me. Unfortunately his own business was going through difficulties. He had started importing tea from the Americas. He showed me samples of excellent quality and said he had negotiated very favorable prices at the source. Unfortunately, however, the import duty canceled out most of the advantage."

He looked at me. "You know that there is an iniquitously heavy Customs levy on tea from outside the Empire, to protect our interests in Ceylon and India?"

"I am well aware of it: it has always been controversial. There was some trouble over the matter in Boston once, I hear." I was trying to lighten the mood, but I got no smile from James.

"Of course, if he could somehow avoid the levy, he would be on to a most profitable thing." He raised his hand. "I will hear no word against McFarlane, mind! He barely hinted at the possibility. It was I who talked him into the thing and pretty much forced him in against his will, at the end. My clinching argument was that my cabs could be used to distribute the tea to shops all around London without attracting attention.

"He had heard of a dockside scam worked by one Lars, a former sea captain living in the East End, with many unwholesome contacts in the seafaring world. Lars knew of corrupt Customs officers who could be persuaded to turn a blind eye to unloading from certain docks at certain times. It was also necessary to pay off the captain of the ship involved, but it was still much cheaper than paying the duty, if a worthwhile amount of tea was involved.

"McFarlane had never attempted the method. For one thing, he had not the nerve. And more fundamentally, he had not the capital: a total of five hundred pounds would be required, and his life savings amounted to only half that. He could hardly have known that this situation presented me with the maximum temptation. For my business account happened to hold just over two hundred and fifty pounds on that very day, although I knew there were several bills waiting to be paid from that. The prospect of tripling the sum overnight hung tantalizingly before me. The profit would put us comfortably back into the black for the indefinite future. And so I withdrew the needed sum from my bank, and browbeat McFarlane into doing the same and taking me to see this Lars."

He had become somewhat hoarse, and I rose to get him another brandy, this time diluting it well with soda from the siphon on the bar. As I sat down again, James looked cautiously around, then continued:

"We reached our destination, a seamen's rooming house at the cheap end of the Whitechapel Road. McFarlane's nerve seemed to be increasingly deserting him, but I asked for Lars, and we were promptly shown into a sparsely furnished back room. A few minutes later, one of the largest and ugliest men I ever wish to see stood blocking the doorway, eyeing us suspiciously. He conversed mainly in monosyllables, and with a strong foreign accent. But I contrived to make myself understood, and it emerged there would be a chance that very evening. If the Customs night-watchman at the East India dock was paid in time, he could arrange matters with the captain of a merchantman that had just docked, and the tea would be placed in a deserted warehouse out of sight of the quayside for collection the next morning.

"The night-watchman would currently be found drinking in a public house called the Blue Anchor. Lars could not be seen there, but if one of us would take the money in an envelope, together with a note that Lars scribbled for us, the scheme would be set in motion. There would be no problem recognizing our man, for he was missing one hand, having suffered an accident in his former job as a dockside winchman.

"'Vun of you goes,' said Lars, 'and vun stays here. Ven de odder come back, I give you orders vor de morning.'

"McFarlane passed the envelope to me. 'You go, James,' he said. 'It is mostly your money, and frankly this business is putting me in a blue funk. I will wait here.' But Lars shook his head. He said that in those surroundings, I would stand out like a sore thumb—he actually used a more vulgar analogy, John—but McFarlane, with a seaman's cap pulled down well on his head, could pass. And so we sent him on the errand, white-faced and trembling with nerves though he was.

"I won't deny that, as the minutes passed with no sign of him, some doubts began to grow in my mind. But my mistrust was ill-founded: presently he came in through the door, not actually whistling, but looking tremendously relieved.

"'It is done,' he said. 'He was in the tavern, just where you told me. The man looked a little baffled when I first put the envelope down on the table. But when I whispered that it was from Lars, he nodded, and the envelope vanished into his pocket like a conjuring trick. If he was not left-handed to begin with, then practice has made him as dexterous as he would originally have been with his right.'

"Then I became aware that Lars was staring at him. 'This man, he has no *right* hand, yes?' he said slowly. McFarlane nodded. Suddenly Lars grabbed his collar and backed him against the wall. 'You fool,' he shouted. 'I told you the man you wanted was missing his *left* hand! You heard me, did you not?'—he turned his angry gaze on me—'I told this idiot it was a man with no *left* hand, who was the vun.' I was in such a state of shock, I found I could not remember for sure. But before I could say anything, Lars released McFarlane's collar, only to box his ears soundly. 'You cretin,' he roared—I am bowdlerizing his words again here, John—'You cretin, you have not only given away your money, you have vith your big mouth made the thing so public that it will not be safe to vork this thing again for many months!'

"McFarlane and I ran back to the pub, heedless now of the need for discretion. But of course by the time we got there, the man he had given the money to was no longer to be seen, nor would any of the drinkers admit to knowing him. You should have seen that pub, John—it made this place look like a palace! It would obviously have been dangerous as well as futile to press our enquiries further."

"Confound McFarlane's incompetence!" I said.

James shook his head despondently. "It is McFarlane you should save your real pity for, John," he said. "Two hundred and fifty pounds is a big sum to me, but I can sell Watson's as a going concern and find some other way to support my children. But McFarlane was gambling his life savings, and more than that, for gangland vengeance is harsh, and I doubt a pair of sore ears will be the last he feels of it. If his blunder turns out to have spoiled their scam for good, he may even pay with his life."

I sighed. "Well, you acted wrongly, but you are a man with responsibilities, and I cannot really blame you very much," I said. "You have been the victim of a bizarre mischance. Imagine two one-armed men frequenting the same pub! The odds on that must be long indeed.

"But," I said, banging my fist on the table for emphasis, "your problems would be solved if you could lay your hands on the one-handed man your partner gave the money to. He will not find it easy to disguise himself."

James shook his head despondently. "If I dared go to the police, it might be done," he said. "But of course I cannot do so without revealing that I was taking part in a conspiracy to bribe one of Her Majesty's Customs inspectors. Which I doubt they would look too kindly upon."

"I am not referring to the official police. I am referring to my colleague Mr. Sherlock Holmes, London's leading private detective."

James choked on his drink. "Have you gone mad!" he spluttered, then hastily lowered his voice as several people looked in our direction. "From what you have told me, he works so closely with the police that he might as well be one of them. Why don't you suggest I simply walk into Vine Street police station down the road and offer my hands to be cuffed? At least it would save some time."

"You are wrong, James. I have known Sherlock Holmes a long time. He is a great deal more sympathetic than you might suppose. I will not repeat your story to him, for I gave you my word. But if you have any sense, you will come to see him this evening—he is expected back at 6 o'clock—and ask his advice."

I duly said nothing to Sherlock Holmes, except that a cousin of mine had done something foolish and that I had urged him to seek my colleague's advice. In fact I was quite surprised when Cousin James did indeed present himself a little after six. I performed the introductions, and after a little hesitation he launched into the same story he had told me. Sherlock Holmes listened with interest until he reached the point where McFarlane was sent off to hand over the envelope. Then he interrupted.

"And when you later found the money had been handed over to the wrong man, what did you do?"

James looked accusingly at me. "You gave me your word you would keep the matter in confidence!" he said.

Sherlock Holmes raised his hand, forestalling my angry reply.

"Watson has told me nothing. I anticipated the development because your story is familiar to me."

"Familiar to you! How could it be, unless John has told you?"

"Not in its specifics, but in its overall form. You have fallen victim to a confidence trick. They are commoner in America, where law enforcement is as yet patchy, but they do occur over here also. In American parlance, you have fallen for a 'Big Con,' an elaborate confidence trick that follows a classic pattern. There are initially two crooks involved, called respectively the 'roper' and the 'insideman.' The roper's job is to approach the 'mark,' befriend him, gain his confidence, and sound him out as a potential victim. He is then introduced to the insideman, who plays a different role: he seems a harder man, and the roper pretends to be afraid of him and makes sure the mark is afraid as well."

"It reminds me a little of the 'good cop, bad cop' play-acting in police interrogations," I said.

"This is a much more subtle game. By appearing hostile to one another, the criminals easily convince the mark that their fabricated business opportunity is genuine; it does not even occur to him that the two might be in cahoots. And so he is persuaded to part with his money. But the real art of the game is in the 'blow-off.' That is the process by which the mark is persuaded neither to report the matter to the police, nor to come after the criminals for personal vengeance.

"There are three elements to the perfect blow-off. First, the mark must be convinced that the roper lost even more heavily than he did, so his feelings toward the latter are sympathetic rather than vengeful. Second, the mark must remain sufficiently intimidated by the insideman that he does not try to pursue him either. And third, the mark must be under the impression that he has himself been involved in criminality, so that he does not dare go to the police. McFarlane and Lars played their roles to perfection. And so, if I may say so, did you."

James was obviously struggling to come to terms with these revelations. His mouth worked but no words came. "And the one-armed men?" he eventually managed to say.

"Did not exist, of course. You saw only McFarlane and Lars."

"Then, is there any hope of recovering my money, now that I know?"

Sherlock Holmes smiled. "There is an excellent chance of it. Pursuing the men would be futile—those will not be their real names, of course. But I would lay a substantial wager you will see McFarlane again. For there is another aspect to the blow-off. If it goes perfectly, the mark may remain so thoroughly deceived that he is actually approached a second time and offered an opportunity to try the game again. You have evidently been an excellent mark, and am sure that you will shortly meet a repentant McFarlane, who will tell you he now has a more foolproof scam, and that he feels a noble obligation to help you make good your losses. We shall then devise a trap for him."

My cousin rose with a sigh of relief. "I cannot thank you enough, Mr. Holmes. I have been very foolish, and if I am not out of the woods yet, I have perhaps come out of it better than I deserved to."

My friend raised his hand. "I am not quite finished yet. There is one aspect of your story that surprises me."

"One aspect! I find every stage of the crime bizarre."

"I was not referring to the crime; that was all predictable to me from your first sentence. But you became tempted by the crime only because your adoption of modern American management methods hindered your business instead of improving it. That I find strange. Tell me a little more about these methods."

James nodded. "I read several books, and although their advice differed in detail, three guiding principles or morals stood out clearly:

"First, targets! Set targets, and put pressure on your employees to meet them. That is the only way to keep people up to the mark. Without targets, everyone works at half-steam.

"Second, margins! Calculate your profit margins, and be aware that a small alteration in costs can make a big difference to profits. You must work out the percentages. Suppose that it costs you 10 pence to make widgets, and you can sell them for 11. If you can reduce your costs by 10 percent, from 10 pence to 9, then your profits increase by 100 percent, from 1 penny to 2.

"Third, activities! Calculate the profit separately on each activity your business does. Everything should make a profit. Otherwise, profitable activities may end up merely subsidizing loss-making ones."

Holmes nodded. "That is all quite sound, as far as it goes," he said. "Now give me examples of how you put these principles into practice."

"As regards targets, my father had been content for his drivers to put in the required hours each day, however much or little they made. I took a cab out myself on several occasions to get an idea of what daily takings should be. My success varied considerably, but the average was 30 shillings if I was energetic in seeking fares. So I abandoned the idea of a fixed

8-hour working day, and each morning I told the drivers they could return only when they had earned 30 shillings. 'Work hard and knock off early, or slack and work longer, it is your choice,' I said. But the success of the scheme is questionable. Some days everyone returns early with the money, and I endure the frustrating sight of the cabs sitting idle in the stable-blocks before the evening rush has even started. And some days the drivers return bad-tempered at midnight, still short of their targets."

Holmes groaned. "That is actually called the *cab-driver's fallacy*," he said. "The self-employed cab drivers are equally prone to it. They set themselves a target for the day, and knock off when they achieve it. The problem is that, as everyone knows, demand for cabs varies greatly from day to day, depending on the weather and other variables. Some days you cannot get a cab for love or money, and other days the streets are fairly jammed with empty hansoms seeking nonexistent fares. It would be far more profitable to follow a policy opposite to that of a fixed target. That is to say, the faster they are earning, the longer they should stay on the street. If business is slow, on the other hand, they might as well knock off for the day."

"It is quite ironic. I suppose they dare not face their wives, to tell them they are taking the afternoon off because they have not earned very much!" I said.

"There may well be some truth in that, Watson. Perhaps what this town needs is more bachelor cab drivers."

James shook his head. "It certainly needs a clearer-thinking cab firm owner," he said. "Anyway, I next turned my attention to margins. For a cab firm, margins are very much a function of where the business is based. The big overheads are rent of the stables and the cost of horse feed. Both vary considerably from district to district. The business had always been based in the East End, where rents are cheap but horse feed is expensive. I realized that by moving to Hampstead, where there is free grazing for horses on the Heath, we could reduce feed costs by half! The rent would be higher, but only by 20 percent."

Holmes frowned. "Yes, but how much was the original feed bill and rent?" he asked.

"In the East End we paid a hundred pounds a year in rent, and horse feed cost twenty pounds a year. In Hampstead we pay a hundred and twenty rent, but only ten pounds a year on feed."

"So your total overhead has gone up from one hundred and twenty to one hundred and thirty pounds. Hardly an improvement!"

James blushed. "Well, when you put it like that . . . But it did seem a waste, paying good money for grass that could be had for free on a common!"

"You have been guilty of being *penny-wise, pound-foolish*, as the old saying goes. Overall, it is absolute savings rather than percentage savings that should be the priority. Innumerable people make that mistake in their everyday lives. Please continue."

"As regards activities, in addition to our core business I had started some other schemes. There was the guaranteed station service, where a cab booked in advance would take you to catch your train at exactly the right time. And the country excursion service, taking couples out into the countryside for a picnic. The station service was very successful, but the picnic excursions never really took off. However, the station service had cost little to set up, whereas we had spent heavily on the picnics, for wicker baskets and fancy cutlery and such. Mindful of the need to make a profit in each area, I stopped advertising the station service and spent all the money on promoting the picnic idea. Our picnic business did increase, but not by very much."

Holmes nodded. "That is the *prior investment fallacy*," he said. "The notion that in a desperate attempt to recoup, you should throw good money after bad. Actually, the logical thing to do is to proceed without regard to previous history. However much or little money it cost to develop a product, that money is spent, for good or ill. The question is whether it now makes a day-to-day profit. You should concentrate on that which sells; the past spending is irrelevant. In this case you should discontinue the picnics, write off the loss, and reinstate this excellent idea of a reliable station service."

James sighed deeply, and rose. "I am most indebted to you, Mr. Holmes—and to you, John. I shall put your advice into practice, and I hope that when we next meet, it is in happier circumstances."

When he had left, I turned to Holmes. "I always thought these American business methods were pretty good. They seem to be rather overrated," I said.

"Far from it, Watson. The business tips he described—attending to cost margins, ensuring that each of your business activities is profitable, and setting targets—are vital to the success of any company. But the knowledge must be applied in the right context. Saving money is good, but not if saving small sums in one area increases costs in another. Unprofitable activities must not be tolerated—but if they cannot be improved, they should be axed, even if you are left with a loss on that venture. Targets are vital, but they must be set wisely so that they encourage useful work, not an alternation between slacking and the futile expenditure of effort."

"I suppose large corporations tend to get these things right?"

"Curiously enough, no, Watson. The errors your cousin made tend to affect both the smallest businesses and the largest. Very big businesses encounter problems of delegation: the interests of their individual departments and managers are not the same as the interests of the company as a whole. For example, they might have a director in charge of office equipment who buys cheap machines that cause minor injuries, because he has no responsibility for time lost through sick leave. They might have a director who insists on continuing a research and development program that he has long known to be futile, because the day its failure is acknowledged will be the day he is fired. They might set targets that are unrealistic—either too high or not high enough—because the head office is out of touch with local opportunities and problems."

"Well, thank goodness I have my common sense, Holmes. I am not as ambitious as cousin James, let alone these big American tycoons, but at least I can conduct my modest affairs without making such blunders!"

McFarlane did indeed contact James again, just as Sherlock Holmes had predicted, and we duly set up a most ingenious trap for him and his confederate Lars. The day the trap was to be sprung, Holmes came down for breakfast early by his standards, and raised his eyebrows to see me with my morning's mail already opened and read, and hastily addressing an envelope of my own in reply.

"Good morning, Watson. What can have brought you to such a pitch of enthusiasm, so early in the day?"

"I have just received a most extraordinary offer, Holmes. It is from a builder who is offering to replace all the doors and windows at my surgery, free of charge."

"Extraordinary indeed, Watson! Does it not occur to you to ask yourself why he is being so generous?"

"It is all explained in his letter, Holmes. He says that having his handiwork on display in prestigious houses visited by the public, such as doctors' surgeries, will be an effective advertisement for his business. But he can only make good his offer on the first replies to his letter, which is why I have signed the contract so promptly."

"Curiouser and curiouser. Your surgery is in a pleasant and clean little house, but 'prestigious public building' is laying it on a little thick. May I glance at this contract before you seal the envelope?"

I passed it across rather reluctantly. Holmes read rapidly. When it came to the small print at the bottom, though, he was forced to pick up his magnifying glass to make out the words.

"Did you read the whole of this contract before signing it, Watson?"

"Of course I did, Holmes."

"Then we must write to the Pope immediately!"

"To the Vatican? Whatever for?"

"To report a miracle, Watson! Your eyesight is imperfect compared to mine, yet I could not read the small print here without my magnifier. However, the magnifier was exactly where I left it last night, with papers partly overlapping it on the desk. I can only conclude that your eyesight has benefited from divine intervention." And with that, before I could stop him, Holmes ripped the contract in two and tossed it into the fireplace.

"You are not going to thank me, Watson? I assure you, I have saved you a significant amount of money. This arrangement would in the long run have cost you more than a reputable builder would have charged for the same service. And in fact there is absolutely no reason why your surgery doors and windows need replacing; a window cleaner and a painter could quickly make them good as new."

He sighed deeply. "I very much fear that this kind of thing is the new version of the Big Con. At least those who suffered at the hands of the old-style gentlemen confidence tricksters were relatively few and, more often than not, were rich people who could well afford the loss. I have a feeling, Watson, that these new Small Cons—letters deceiving millions of ordinary people into purchasing apparent bargains—are going to grow in number until they pull in far more money in total than the old Big Cons ever did, and from poorer people who can ill afford it." He shook his head reprovingly.

I felt it was time to move the conversation on to a new subject. "Would ham sandwiches suit you for lunch, Holmes?"

"Ham sandwiches? Of course, Mrs. Hudson is away. Yes, that would do me most admirably. And if you are passing the local tobacconist, perhaps you could get me a quarter of Old Sailors'? We have enough for another day or two, but it would not do to run out. Make sure you are not gone long, though. There is no telling when McFarlane will come, and if he brings Lars I will certainly need your assistance."

However, it was nearly an hour before I finally returned with the groceries. I was relieved to find Sherlock Holmes still alone, keeping watch from the window with his spyglass.

"You seem to have made rather heavy weather of things, Watson!"

I spoke with some heat. "The local shop was out of your tobacco—they are expecting a fresh delivery tomorrow. It is not such a common smoke: I had to try half a dozen shops before I found it."

"I hardly meant to put you to so much trouble. I told you we had not yet run out. Why did you not simply wait for the local shop tomorrow?"

"Well, you know me, Holmes. When I have put my mind to accomplishing a task, I do not like to return with it not done. I put in as much effort as it takes."

"The cab driver's fallacy," Holmes murmured. "And did you get the edibles at the south or the north end of Baker Street?" (There was at that time a little cluster of shops, each including a baker and butcher, at either end of the street.)

"I went to the south end. The bakery there sells loaves for a penny; at the north end they are tuppence. Quite monstrously overpriced: double the money for the same loaf, it is a scandal!"

"Yet the butcher at the south end is the dearer, is it not?"

"Well, yes, but the ham only costs a quarter more than at the other place: I paid 15 pence rather than 12."

"So our total lunch cost 16 pence, whereas had you gone the other way, it would have been 14 pence. Curious economics, Watson. In fact, it is the penny-wise, pound-foolish error."

"Well, I simply refuse to be so extravagant as to pay double price for bread!"

"Actually, I noticed you at the south end. I have been observing the street for some time, looking for a face more sinister than yours, I need hardly add. And yet I noticed a curious thing. When you came out of the baker's, before going into the butcher's, you passed Milly the sandwich lady.

"Now I could see that she had unsold sandwiches still on her tray. I know that if she has any left over after the morning rush, she sells them off for a penny each, because otherwise they would be thrown away. Without disparaging your sandwich-making abilities, hers are extraordinary, as I am sure you know. Why did you not abandon your quest for ham and buy better sandwiches more cheaply, also saving yourself trouble?"

"But I had already bought the bread. It would have been stale by tomorrow, and would have gone to waste. There are children starving in Africa, Holmes!" I waved the wholesome bread loaf at him. "I really think it is a crime to waste food under those circumstances, even if it doing so would save me a few pence and a little trouble."

"You have not helped them by donning a hair shirt and sentencing us both to your rough-and-ready sandwich making in place of Milly's delicate fare. Whereas if you had bought her sandwiches, thrown away the loaf, and put the pennies saved in the missionary box, it would have accomplished something positive. You have committed the prior investment fallacy.

"I really must congratulate you, Watson. In the course of one morning's ordinary domestic decisions, you have managed to replicate on a small scale every one of the errors that brought your cousin's business to its knees! Come, do not pull such a long face. Everything improves with practice, and I am sure you will get plenty of that. I believe Mrs. Hudson is away for the whole week."

2

The Case of the Gambling Nobleman

SNOWFLAKES TUMBLED RANDOMLY PAST THE windows of our Baker Street apartment. But I stood close to the glass, grateful for the cold, for my colleague had stoked the fire until the air in the room was more than comfortably warm. Sherlock Holmes took his pipe from his mouth and regarded me quizzically.

"A New Year's resolution, Watson? You have never asked my advice on such a trivial matter before."

I indicated the calendar, where only the last day of 1899 remained to be crossed off.

"This is no ordinary new year, Holmes. It is an opportunity to make a resolve for the whole of the coming century!"

My friend smiled. "You feel that a conventional decision to eat less cheese, or perhaps take a little exercise before break-fast"—here he regarded my waistline closely—"does not quite fit the bill? Hmm, well, perhaps something a little more ambitious *is* called for." He hesitated. "May I speak frankly, Watson?"

"My dear fellow, of course you may!"

"Well then, this is what I would recommend. If you could only be a little less—that is to say, we all have the limita-

tions of the brain we are born with, but you could still perhaps . . . What I am trying to say, Watson, is that if you
would only take time to think more clearly, resolve to be a
little more logical, a little more scientific in your everyday
decisions, you would be a happier and more prosperous
man for it."

I flushed hotly. "Really," I exclaimed, "I would be the first
person to admit that as regards specialized tasks such as detection, I have my limitations. But when it comes to running my
own life, sound common sense serves me quite well!"

Holmes shook his head. "If you could manage to improve
the odds just a little in the gambles you make every day—"

"I never gamble, Holmes!"

He smiled. "But you do, Watson: it is a truism that life is but
a series of gambles."

Sometimes there is no point arguing with Holmes. My common sense told me that it would be wise to change the subject, before we marred New Year's Eve with an unseemly row.

"Can I at least persuade you to stay up and toast the New
Year with me?" I asked.

"No, Watson, I see no reason to be excited by the utterly
predictable appearance of a new number upon the calendar. I
am planning an early bed, and have no intention of venturing
out before the morrow. Given the volume of noise and
drunken revelry to be expected, I judge it a particularly good
evening for staying in."

"Well, I deplore your cynicism, on a night when all the rest
of the world will be out enjoying itself!"

Holmes shook his head. "Every year the Christmas and New
Year festivities stretch longer, and for many they are far from
enjoyable. Look out at the faces of those in the street below,
and tell me what you see."

I looked down. Certainly there were few happy expressions
to be noticed.

"That policeman, now. Did you ever see a more hangdog
face?" he asked.

I looked at Holmes in astonishment. He was still in his arm-chair by our blazing fire, and could not possibly have looked through the window to the street below.

He smiled. "It was an easy guess that in the throng, there would be at least one policeman visible nearby. And under the new shift system, if he is on duty tonight, then he will also have been pounding his beat on Christmas Day. That is perennially the least popular shift of the year."

"He cannot still be mourning a late Christmas dinner!"

"No, Watson, the reason is less trivial. Nowadays we are all supposed to be happy and convivial at Christmas, as a result deepening the misery of those perforce alone. For some, only one solution appears possible. The constable you see will almost certainly have had to assist in the cutting down of at least one such unfortunate."

He warmed to his theme. "And I fear that tomorrow a similar story may unfold. For just as we are wished a happy Christmas, so we are wished a prosperous New Year. At year end, many a man sits contemplating the paucity of his worldly achievements, and the magnitude of his debts, until the dawn brings not hope but despair."

I was determined not to have my excitement at the new century spoiled by my friend's black mood. "Well, I suppose that in general those who find themselves with so little have only their own idleness or folly to thank," I said.

To my surprise, Holmes shook his head. "Once I might have said so, Watson. But the more I see of life, the more I am struck by how great a part is played not by brains, or skill, or character, but by simple chance. The success or failure of a business venture, or a marriage, or a war, can depend more on the blind roll of Fate's dice than on any planning. Life is a chaotic business, and the most unpredictable of happenings can determine the fate of one man's life or of a whole nation. The whim of Lady Luck rules all.

"Naturally, successful men like to credit their own skill, rather than mere chance. The rich stockbroker, passing the beggar on the pavement, congratulates himself on his skill in getting ahead in life. But perhaps the broker had no more real ability to predict the market than the beggar, and blind luck chose who was to dine on champagne and caviar, and who on slops." He paused to stuff his pipe. I did not know what had led to this unusual outburst, but I could see he was a little embarrassed, and evidently felt that he had overstated his case somewhat.

"Of course, that does not mean one cannot maximize one's chances by the use of intelligence. Life is inevitably a series of gambles, but knowing when and in what circumstances to take a chance is vital. Napoleon, that greatest of generals, notoriously asked a curious question when he wanted to find out whether a subordinate officer was worthy of promotion. He would not ask about the man's tactical skill or strategic understanding. He would ask merely, 'Is he lucky?' For he knew that in order to be perceived as lucky, the officer must have become expert in calculating the odds of the gambles he must take, which is the most vital skill on the battlefield, just as it is as in life." He paused. The church bells had ceased their ringing, and in the momentary hush, the noise of drunken quarreling rose clearly from the street below.

"Perhaps you are right, Holmes, and this is no night to be abroad," I said.

"Well spoken, Watson! If some risks in life are inevitable, some are certainly avoidable, and tonight, with a good pipe and a bottle of vintage port close at hand, there is no power on Earth that will move me from this chair."

At that moment there came a firm knock at the door. I opened it to reveal not a man, as I had expected, but a soberly dressed young woman of wholesome rather than beautiful appearance. Her confident expression and the firm set of her jaw were nevertheless more impressive than mere prettiness. She looked me up and down and then, without invitation, pushed past me into the room and addressed my friend.

"Mr. Holmes, they tell me you are the cleverest man in London. I have come to ask you if you will stoop to giving some advice to the man who may well be the most foolish!" She spoke with a trace of an American accent.

My friend's lips barely twitched, although I could see that he was pleased by her words. He gestured her to the chair opposite him, waited while she pulled off her gloves, and then rose to take her hand, bowing over her as he did so. "You flatter me, madam. I know of at least three men comparable in intellect to myself, and one definitely superior, who live within this metropolis. Let us hope that your opinion of your moneyed but unreliable fiancé is similarly exaggerated."

The lady hesitated. Holmes smiled at her. "A young lady wearing an engagement ring, who comes to me concerned about a man, is likely to be speaking of her fiancé. And a man who buys his intended a diamond-studded ring has money. But if the ring has shortly to be pawned—I saw the broker's mark still on the metal—his dependability is perhaps less to be trusted."

"I see that I am more obvious then I thought, Mr. Holmes. Let me introduce myself: my name is Catherine Lawrence."

"And how is it that you think I may be able to help you?"

"My fiancé is the Marquis of Whitebridge. You have probably seen his name on the social pages of the newspapers from time to time. But while he might seem to lack nothing, his life is lived under something of a cloud. For although he will inherit a noble title, the fortune that should accompany it is much depleted. His father and grandfather were both less than wise in their investments, and now the capital that goes with the estate is barely sufficient to cover its expenses.

"I assure you, that means nothing to me," our guest sighed. "But to Lionel it is a humiliation. He feels that he cannot marry me until his future is more assured. And rather than take a job, which I would encourage, he believes that the only route to true security is to find some way to increase his capital. To this end, he has, like his forebears, engaged in several risky business ventures. One or two have been quite successful—that is how I was enable to redeem my ring from pawn, as you observed. But most have been failures, and where once his capital might just have kept us in adequate comfort, it is now definitely insufficient."

Here our visitor shuddered. "As the new year approaches, he has grown more desperate. I suspect he had long ago promised himself that he would restore the decline of this century before the new one dawned. A recent venture frightened me. He studied classical philosophy at Cambridge and has always been convinced that he has a deeper insight into the laws of mathematics than that held by modern students of the subject. Of late he has been obsessed with the notion that he can formulate a gambling system that will enable him to increase his money more rapidly than any conventional investment."

She looked at Sherlock Holmes, but he made no sign or comment, and she continued.

"His first scheme had the merit of simplicity. It is well known, he says, that the laws of probability always assert themselves in the long term, but not in the short. Thus, for example, if you toss a coin two or three times, it may well turn up the same way each time. But if you toss it a few hundred times, you are very likely to get a more even division between heads and tails. As the number of times you toss it tends toward infinity, the ratio of heads to tails is guaranteed to become ever closer to unity."

"That is quite true, so far as it goes."

"Following this principle, Lionel resolved to bet only on the roulette wheel, and then with restraint. He would wait until an almost uninterrupted sequence of one color—either red or black—had continued for some time. Then he would place a modest bet on the opposite color. For he knew that in order for the laws of chance to be obeyed, a preponderance of the other color would become inevitable."

Sherlock Holmes sighed. "And has he already put this plan into effect?"

"Yes, a month ago. He went to the Casino Royal in Piccadilly on several successive nights. He explained the game to me. The wheels have eighteen black and eighteen red numbers, and you can bet by choosing either black or red. If your color does not come up, you lose your stake; if it does, you win an amount equal to it. Of course there is also a thirty-seventh number on the wheel, the colorless number zero, and when *it* occasionally comes up, all stakes are lost. That is how the house makes its percentage."

"And what luck did he have?"

"Well, in a way, less bad than I had feared. I refused to accompany him to the place, but he recorded all his bets meticulously in a notebook, which he showed to me afterward. In all, he placed bets of value equal to four thousand pounds." She shook her head. "Had he lost a sum of that magnitude, he would have wasted most of his remaining inheritance. But in fact he lost only about one hundred pounds, though even that is money he can ill afford. At that point he had the sanity to admit his system was not working, and he gave up."

"To what did he attribute the failure?"

"Why, to that pesky number zero! He told me his scheme did seem to function in principle. After a run of black, a red would indeed turn up before long, usually within a couple of spins. But evidently the house percentage was large enough to swamp out whatever marginal advantage his system produced. As I could have told him, for casinos would hardly stay in business if there were such an easy way to make a profit at their expense."

Holmes spread his hands. "If he has decided for himself that gambling is futile, why then do you need my help?"

Her lips pursed, and she spoke angrily, yet barely above a whisper. "This morning, I called upon Lionel without warning.

I intended to talk to him about resolutions for the New Year in the hope that I could guide him into making some sensible decisions about his behavior in the new century. He was not yet up and dressed when I arrived, and I was shown into the morning room. As I waited there, I noticed a banker's draft upon the side table. A draft for ten thousand pounds—already endorsed to the Casino Royal! That must represent a mortgage upon his entire estate.

Miss Lawrence struggled to maintain her composure. "When he came down I told him what I had seen. I pleaded with him, but in vain. He is adamant: he believes that with a larger sum of money, he can succeed where before he failed. He told me he plans to go to the casino at 8 o'clock tonight and swears that before midnight, he will visit me with his family fortunes restored, to pledge a wedding date as the bells ring in the new century. It is a madness that has come upon him, Mr. Holmes! Almost, I handed him back his ring. But then I thought of coming to you."

Holmes smiled at her. "It is a little out of my usual line of work, but preventing human folly in one way or another is no novelty to me. Have no fear. Watson and I will make our way there shortly, and I am certain that when the Marquis arrives, we will be able to tell him a story that will cure his gambling habit for good!" And with a few more words of reassurance, he escorted her to the door.

I lay back in my chair and roared with laughter. "What folly, Holmes!"

"You do not think his gambling system sound, then?"

"Obviously not. How could a roulette wheel possibly remember which color had come up previously and hence favor the opposite on the next spin? It is the same machine each time, so the colors must come up with the same probability on each turn, regardless of the previous outcome."

"That is sound common sense, Watson. However, you would not deny the lady's assertion that the laws of chance assert themselves in the long run? Spin the wheel ten times, and you might well get, say, seven reds to three blacks. But spin it ten thousand times, and the ratio of red to black would in general come out much closer to equal."

"I certainly believe that is what the mathematicians say."

"Then answer me this, Watson. You embark on this marathon ten-thousand-spin test. After ten spins, you have recorded seven red and three black. You know that the ratio will in due course return toward unity. *But how can that happen, unless some subtle mechanism is now encouraging the wheel to turn up black slightly more often?*"

"Why, I—I cannot imagine, Holmes! Obviously, to restore the numbers so that the black count more nearly equals the red, some corrective force must operate."

"That is quite true," Holmes said quietly.

"Then I can only suppose there are indeed some as yet undiscovered subtleties in the laws of Nature. Perhaps in some holistic way, the universe retains a memory of previous events, to correct the balance of probability in due course?"

Holmes brought his fist down on the window ledge with a thud that made me jump. "Nonsense, Watson. You did not listen properly. I did not say that the *numbers* of red and black will become more nearly equal. I said that the *ratio* of red to black would do so."

"Well, if there is a difference, I am afraid it escapes me."

Holmes sighed. "It is easily demonstrated with algebra, Watson—no, have no fear, I will respect your allergy to mathematics." He turned back to the window. Suddenly, he gave a snort of laughter. "Now, there is a coincidence if you like, Watson. Outside there, I see the ideal man to explain the point to you, and without using so much as one algebraic symbol!"

I hurried to the window. The blizzard had evidently driven the earlier crowd indoors, leaving the street deserted, although only a few flakes were still falling. A layer of snow covered the ground, concealing the litter and gutters beneath a blanket of white and giving the view a pure Christmas-card beauty. Regrettably, this was marred by a zigzag line of footprints down the center of the road. A very drunk sailor—doubtless a liberty-man from the ships moored at Greenwich—was slowly attempting to make his way northward. However after every step he took forward, he could not help lurching sideways either to the left or to the right. There was no one else in sight. I looked at my friend in bafflement, but he nodded.

"Yes, Watson, I am referring to the gentleman in marine uniform."

"Really, Holmes, that sot is obviously incompetent to explain anything to me!"

"Not in words, Watson, I grant you. But his movements are most eloquent. A fine demonstration of the Drunkard's Walk!"

Holmes grasped a sheet of paper from the table and drew a sketch to illustrate.

The Drunkard's Walk

"That is just the name of a well-known technique for graphing a random series. Suppose that a man starts from a point in front of you and attempts to walk north, along the dotted line. However, after every step he tosses a coin and then takes an additional side step left or right, depending on whether the coin falls heads or tails. Thus, he progresses in a series of diagonals, like the man we saw below, and his track records the sequence of heads and tails. At a given moment, we can easily tell from his position whether the heads have exceeded the tails or vice versa. If he is standing on the dotted line, the number of heads has exactly equaled the number of tails at that point; if he is three paces to the left of the dotted line, there have been three more heads than tails."

Holmes's eyes narrowed. "Now, we shall assume the man cannot even see the line. There is no certainly no mystic force drawing him back toward it! As you might expect, the farther he walks north, the farther he is likely to stray from the line. Nevertheless, from our point of view, the farther he walks, the more nearly he will come to being exactly northward of our position. This is because the *ratio* between his distance north of us and the distance he has veered to east or west is likely to diminish. So by the time he is but a dot on the horizon, we will see him very close to due north."

"I think I grasp what you are saying, Holmes. But I am still not sure I understand *why* the ratio should diminish."

"Well, consider for example a moment when the man is ten steps to the west of the dotted line. When he takes the next step, will his distance to the east or west of you have increased or decreased, on the average?"

"He is just as likely to veer east as west, so the average change will be none."

"And yet his distance *north* of you will definitely have increased by one pace. So—on the average—the ratio of his sideways distance to his northward distance from you will be just a little bit reduced. Of course there is one counter-example: if he is actually on the dotted line, a step either way will take him farther from it. But nevertheless, the farther he goes, the less the ratio of his total sideways to his total forward progress is likely to be, because the forward motion is always cumulative, and the sideways motion often cancels.

"Suppose that he starts out, by chance, with three consecutive paces all to the left. That starting bias will never be consciously corrected: a thousand paces later, he will still be six paces to the left of where he *would* have been, had those first paces been instead to the right, and the rest of the sequence the same. But by the time he has made a thousand paces, and is hull-down on the horizon, the difference those six paces make to the *angle* at which you see him will be negligible. In terms of direction, you will see him in the long term converge inevitably upon the northward."

I looked out of the window at the flakes still tumbling past the panes. Below, the deepening snow was steadily covering the features of the street. Already, you could not tell where the road ended and the pavement began.

"I am satisfied, Holmes," I said. "Any advantage of heads over tails, or of red over black on the roulette wheel, is smoothed out in the longer term, not by any kind of active cancellation, but by a more gentle and undirected happening, a kind of washing out. That is, the effect of any flukish run gradually becomes ever more negligible relative to the whole. At last I understand the matter, perfectly and in full!"

Holmes settled back in his chair contentedly. "I am relieved to hear it, Watson. I am sure you will do an excellent job of advising the Marquis."

I felt an uneasy feeling grow in the pit of my stomach. "*I* advise the Marquis, Holmes? But I assumed you were planning to go yourself."

"I feel that my presence would now be superfluous, Watson. It is such a relief to be able to delegate a tricky task to your dependable self. I have always had the highest trust in you, but nowadays, you positively astonish me with your intellectual abilities also. Let me pour you a whiskey. It is getting to be quite a foul night, and the internal warmth will sustain you. There are a few minutes yet before you will need to leave."

As I sipped my whiskey, I felt a glow of satisfaction at the trust my partner was placing in me. I was confident I could carry out the task delegated to me. And yet—I could still feel some little seed of doubt, some unconscious reservation, gnawing away in my mind. Could it be the Marquis's claim that his system did seem to have been partially working, notwithstanding the spoiling factor of the zero on the wheel? He claimed that after a run of black, red tended to show up within a spin or two. But that was hardly surprising. After all, at any point in time, the chance that red will not show up in the next two spins is always only one-quarter: one-half multiplied by one-half. Or only a little less than that, at any rate, allowing for the zero. He had bet four thousand in small sums and lost only one hundred or so. But that was also right: the zero would come up about one time in thirty-seven, and the luck of red and black would tend to cancel the rest of the time.

It was a much older memory that troubled me. My eye was drawn to the bookcase on the far side of the room where my medical textbooks and also my old schoolbooks reposed. It took only a few seconds to locate the page I dimly recalled.

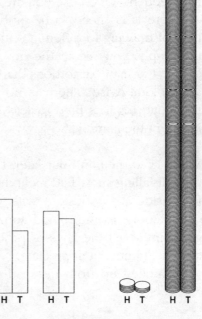

Heads Versus Tails

"See here, Holmes. You claimed that in a long run of coin tosses, it is only the *ratio* of heads to tails that tends toward unity, and that the *absolute* discrepancy between the number of heads and tails could only tend to increase."

"So I did, Watson."

"But here is a bar chart that contradicts you. It shows the outcome of an actual experiment in which a coin was tossed ten times, then five hundred times. And the bars on the right are much more nearly equal than those on the left."

My friend laughed. "Not so, Watson. You are overlooking the elementary fact that the bars on left and right are drawn at different scales. The text mentions that in the first trial, the coin came up 6 heads to 4 tails; in the second, 262 heads to 238 tails. Let me sketch the actual piles of coins in question. Now you can see it: the absolute difference between the piles on the right is much greater."

He drew in the margin to the right of the bar graph. I felt my face redden. "Of course you are right, Holmes."

"There is no need to be embarrassed, Watson. The simple trick of drawing to different scales often makes a nonsense of the graphs you see in the newspapers, and this is only the simplest of many distortions that may be introduced, intentionally or otherwise. There is no further doubt in your mind? Excellent, for it is time to leave for the casino if you are to intercept the Marquis."

When I returned an hour later, I was both cold and damp, for the still-falling snow had melted upon my boots and greatcoat. But inside I was flushed with pride: in return for Holmes's explanations to me, I had some remarkable information to give him about the laws of probability. Something I was certain he did not yet appreciate.

He looked up from his reading. "Was your mission a success?"

"Entirely, Holmes. The Marquis accepts completely that the probability of a random event, be it a coin toss or the spin of a roulette wheel, will be quite unaffected by the result of any previous games, and he will never again attempt to use such a system."

"Bravo, Watson! There was no unpleasantness, then?"

"On the contrary, the Marquis was most cordial. He insisted on buying me a further whiskey at the club bar, by way of thanking me for my concern."

"That explains the rosy glow of your cheeks! And I take it he then left the casino with you, to visit his fiancée and wish her a happy New Year?"

"Well, let me tell you how it went, Holmes. The casino was not at all what I was expecting—I suppose I was naively imagining something like a betting-shop but on a larger scale. On the contrary, the main hall at the Royal is the grandest room I have ever been in. Imagine the ballroom at the Ritz, but larger and brilliantly lit by electric chandeliers. The bar is a gallery around the side, so we could see everything that took place. The tables, the croupiers, the uniformed guards smart as soldiers on parade—it is spectacular. And the gamers! I saw Lord L—, and Lady C—, and quite a few others I recognized from the Court pages of the newspapers but could not name. Not to mention the industrialist Sir M—. Clever, successful people, Holmes. Surely their like would not go there if nothing was to be won by gaming?"

Holmes smiled. "If you are very rich, a trip to the casino may cost no more than a theater ticket to the likes of us. And you usually lose only a small fraction of what you gamble, so you can impress a lady with the depth of your purse while paying but a percentage of its contents. However, it is not a sensible pastime for the likes of you or me, or indeed our client. I assume he *did* come out with you, Watson, for I notice you have not yet answered my question directly."

"He did not, Holmes. For he has discovered something quite remarkable. Accepting that the laws of chance have no memory, he has nevertheless devised a way to win systematically at roulette! Modest amounts at a time, to be sure, but without any possibility of loss."

My friend spoke abruptly. "Stop, Watson; do not remove your boots. I feel the need to join you in a little constitutional walk."

He threw on his coat and hat, and shortly we were strolling south down Baker Street, in the direction I had just come from—toward Piccadilly, home of the Casino Royal.

"I quite understand, Holmes," I said, "that you are curious to see this remarkable place and observe the Marquis's system in action."

Holmes sighed. "Watson, there is no honest way to gain consistently at roulette. It is possible to win in the short term, but in the long run, no system works. The house gets on average 3 percent or so of each bet, quite irrespective of the choices made by the gambler. But do not blame yourself. I should have anticipated this development, and any money the Marquis loses while we are en route will doubtless help to underline the lesson I am about to give him. Tell me about this new system of his."

"It is beautiful in its simplicity, Holmes. First the Marquis bets one pound on the red. If red comes up, he wins a pound. If red does not come up, he doubles his bet on the next spin, placing two pounds on red. If he wins this time, he has lost a pound but gained two; again, he is a pound ahead. If red *still* does not come up, why, he just doubles his bet again, and if he wins four pounds, then, having previously lost three, he is again ahead exactly one pound. And so on, doubling up each time. The system is quite infallible, because by the laws of chance, red must come up sooner or later, and at the close of the sequence, long or short, he will always have won one pound. Then he starts a new sequence. It is an industrious method of gambling, winning just a little each time, but the elimination of risk makes it well worth it."

As I was talking, my companion had been lengthening his stride. We swung out into the street to avoid crowds of revelers who were beginning to pour out of the pubs, merry and singing, and Holmes picked up his pace even more.

"Faster, Watson! Far from eliminating risk, the Marquis is risking his entire fortune each and every time he starts such a sequence. The risk is rather slight each time. But if disaster strikes, it will be total. Usually the Marquis will indeed win a pound. But he has taken ten thousand with him. Tell me, what will happen if the wheel turns up black or zero thirteen times in succession?"

I calculated in my head. "After one loss, he needs to bet two pounds. After three losses, four. Four losses, eight. After ten losses—good gracious, it rises rather steeply—he will need to bet just over one thousand pounds to recoup. After thirteen losses it will be over eight thousand."

"And how much will he have left by then"

"Well, he will already have lost one plus two plus four plus eight . . . why, of course, it adds to over eight thousand, being just one pound less than the fresh bet."

"And since he started with only ten thousand, he will not have the resources to place the bet, and his fortune will be lost!"

"I can see that is a theoretical possibility, Holmes. But surely the chance of thirteen blacks in a row is very slight?"

"A corresponding one chance in eight thousand. Each time he starts a sequence, he is merely setting a high probability of winning one measly pound against a lower probability of losing eight thousand. The odds still balance. And furthermore, we must take the zero on the wheel into account. The chance of losing his stake each time is actually not 1/2 but 19/37. Over thirteen runs, the difference is significant: with each new sequence, he is actually taking more like one chance in six thousand of disaster. As ever, the odds favor the house."

"I suppose to win safely with the system, you would need very large resources?" I gasped, short of breath as I struggled to keep up with my friend.

Holmes snorted. "Not merely large, but infinite!" he said. "Even if you had a million pounds, although the chance of a loss would be very slight, the magnitude of the loss when it eventually came would be even more devastating in proportion. In fact, being a millionaire would not help in practice, for neither does a casino have unlimited resources. Accordingly, there is always a house limit, a maximum amount you are allowed to bet. In practice, I doubt you would find a wheel on which single bets over a few hundred were permitted, even in Monte Carlo."

I had thought we would need to go up to the gallery bar to locate Whitebridge, but it was not necessary. There was an admiring crowd around one of the tables: Holmes elbowed his way roughly through, and the Marquis came in sight at its center. He already had a substantial heap of chips before him and a look of great confidence on his face. His expression changed somewhat when he saw us, and I introduced my friend.

"Delighted to meet you, I am sure, Mr. Holmes," he said perfunctorily. "But as you can see my system is quite sound. I have already won forty-two pounds, and progress is reassuringly steady."

Sherlock Holmes bowed. "I see that your system has triumphed. However, I feel your fiancée may be displeased if I cannot tell her you permitted a consultation, at her request."

Whitebridge looked sullen but allowed himself to be led aside. My friend quietly outlined the explanation he had given me. I saw the Marquis's face suddenly blanch. Presently he went back to the table and gathered up his chips. Ignoring the disappointment of his audience, he took them over to the cashier's desk to exchange for crisp new notes, and we left the club together.

"I suppose I must thank you for saving me, Mr. Holmes. I will go now to give Catherine the news."

My friend raised a stern finger. "One moment. You forget my fee. Allowing for being called out on short notice at such an hour, on such a day, and in foul weather to boot—why, it comes to just forty-two pounds."

The Marquis paid with bad grace, and we took our leave and started the walk home. I noticed that the crowds that had been emerging earlier were now gone from the streets; in fact all was silence, save for a distant splash from the direction opposite to that in which the Marquis had gone. I checked my watch: to my chagrin, it was a quarter to one.

"We have missed the new century, Holmes," I said. "But I must confess it does not really feel so different from the old one. How annoying, though, to miss the stroke of midnight in order to rescue a man so foolish he did not realize a roulette wheel has no memory."

Holmes shrugged. "It is an astonishingly common belief, Watson, to think that the laws of chance require outcomes to come back into balance. That a run of bad fortune makes good more likely to follow—or indeed the opposite, that because you have escaped accidents for a while, one must soon be due. Perhaps people unconsciously assume that Fortune has a finite number of outcomes in the sack of black and white pebbles she carries. Then the more black pebbles you are dealt, the higher the proportion of white remain in her sack, and the more likely you are to get white. But in truth her supply is infinite, and she can always continue to give black or white at perfect whim. Failure to understand that is the first great human fallacy in misunderstanding the Laws of Chance."

I snorted. "Well, frankly, my native common sense would have prevented my ever trying the Marquis's first system, even if I did not understand randomness quite so clearly as I thought."

"The second great fallacy is to think that you can ignore a very tiny chance of a very large loss or gain. A mathematician would warn you of the meaninglessness of multiplying zero by infinity, but we did not have to venture into such abstractions to see that the Marquis's second system would have come to grief eventually. And I fear you were at least temporarily taken in by that one, Watson."

My friend took my arm. "But now let us hasten back," he said in a kinder tone. "We have done a worthwhile night's work, and I am after all still awake to greet the New Year with you. There is some fine malt whisky on the mantel, and it is not too late to drink a toast to our good fortune in the new century."

"We will drink also to Lady Luck!" I said solemnly. "For she is more subtle and powerful than I knew."

Later, it was the effect of quite a number of consecutive toasts—after all, a new century does not dawn often!—that must have blunted my usual tact.

"You know, Holmes," I said, "there was something in your conversation with the Marquis that surprised me."

"Really? What did I say that took you so aback?"

"It was not what you said, but rather something you did not say. I remember being surprised by your omission at the time."

"Go on, Watson." His tone was anything but encouraging, but I took my courage in both hands and proceeded.

"You explained very cogently why his particular system must fail. But surely you could have made a much more general statement. You could have told the Marquis that all gambling is foolish, under all circumstances, and can only lead to loss in the long run. A warning strong enough to prevent him from any future temptation. After all, there are many forms of gambling that do not involve a roulette wheel."

"I could not have told him that, Watson."

"But why not?"

"Because it is not true."

"Really, Holmes!"

My friend sighed. "Explain to me, Watson, how you could ever make such a general statement as 'Gambling is never sensible.'"

I saw that I would have to be a little careful in my choice of words. "Well, Holmes, I know that there is a sense in which a stock market investment is a gamble, because share prices can fall as well as rise. And yet I own stocks and shares, and I know that you do as well. That is rather different: someone has a good use for my capital, and for every pound I invest, I expect to get back more than one pound in the long run, which is sensible. But gambling in the sense of betting is surely always unwise. In any casino or bookmakers, the house must take a percentage to pay for the staff wages, the rent on the premises, and so forth. Thus, for each pound gambled, I will on average get back less than a pound. Which is always *not* sensible. So there you have it."

"Watson, talking to you is ever instructive. In one short speech, you have managed to include no fewer than three fallacies. Let me find you a counter-example to each.

"First, you state that you can never find a situation where you can bet a pound in the expectation of getting more than a pound back. That is true of the roulette wheel, but there are many different games to which it does not apply. Even in a casino, for instance, you can make a systematic profit at blackjack if you can perfectly remember what the fall of the cards has been and do the associated calculations in your head. But that feat is beyond nearly all of us, which is why casinos continue to offer blackjack. Moreover, they attempt to spot card counters and ban them from their premises if they win regularly.

"A better example is to be found at the bookies, betting on horse racing. A bookie always sets his odds so that they add up to a little less than unity. For example, in a race between two equal horses, he would offer slightly less than even odds on each—and so gain a slight profit whichever horse won. Bookies adjust their odds continually, shortening the odds on a given horse as more is bet on it, so that whichever horse wins, their profit is approximately the same.

"If all punters were equally knowledgeable and intelligent, there would be no room for profit. But in fact they are not. Punters tend to bet more on long-odds horses, such as hundred-to-one outsiders, than they should. And that tends to make the odds on good horses a little better. There is in particular the phenomenon called the 'housewife's favorite.' When the name of a horse has become very well known in a way that does not reflect its current ability, for example, because it won many races in its youth, or suffered some experience such as being kidnapped and recovered—"

"That was certainly one of your triumphs, Holmes!"

"—then if the horse is entered for a major race such as the Derby, folk ignorant of horse racing tend to bet on it out of sheer sentiment. That quite markedly improves the odds on more promising horses. From the bookies' point of view, they must lengthen the odds on the realistic favorites to encourage more betting on them, and so get enough takings to insure themselves against the (admittedly remote) possibility that the famous horse will win, and they will have to make large payouts on it. When a more obscure horse wins, as usually happens, the bookie makes his normal modest profit, and the housewives' money goes to the shrewder gamblers.

"There are other ways for a professional gambler to stay in business. It is best to find a venue where there is no house percentage, such as a poker game with 'friends.' The professional calculates the odds of each hand more accurately than the amateurs with whom he plays, so he tends to come out ahead.

"So, let us have no claims that you can never profit by gambling! To do so systematically, however, you need a shrewd knowledge both of probability and of the game you are playing, and the likes of the Marquis should certainly steer clear of the occupation."

"I still think it is an immoral way to earn a living, Holmes. After all, the professional gambler's profit is always someone else's loss."

My colleague ignored me. "Second, Watson, let us consider your assertion that you should always gamble if the expected return—that is, the reward on offer divided by the probability of winning that reward—is greater than the stake. In many states in America, there is a lottery for which millions of tickets are sold, each conferring a tiny chance of winning a huge prize. A fixed proportion of the ticket sale receipts, say two-thirds, are allocated to the winner's fund. Thus, if three million one-cent tickets are sold in a given week, the winner scoops twenty thousand dollars."

"Quite so, Holmes, an average return of only sixty-seven cents per dollar invested. Which is why I would never buy such a lottery ticket."

"Ah, but you see Watson, there are some weeks when no ticket wins—no one has chosen the number that comes up. When that happens, the prize fund is 'rolled over': it is added to the prize money for the following week. Hence, if an equal number of tickets are sold the following week, the expected reward is now forty thousand dollars, divided by one chance in three million, which is one and a third cents per cent invested. A better return than any available from dull old stocks and shares! In practice, more tickets are sold in a rollover week, diluting the benefit somewhat. But it is nevertheless quite often the case that the average return is greater than the ticket cost. Would you not feel obliged in those circumstances to spend all your savings on lottery tickets?"

I hesitated. "I suppose if I were a truly logical investor, I would have to," I said.

"Nonsense, Watson! I am sure your common sense would protect you from such folly. The action would be overwhelmingly likely to leave you a pauper. But a mathematician would explain such restraint in terms of what is called 'utility.' That is simply a way of saying that a given sum of money is not of equal value to all individuals in all situations. For example, the gain or loss of a single dollar to an impoverished student may matter more in practical terms than the gain or loss of a thousand dollars to a wealthy man. In general, the more money you have, the less significance to you each additional dollar has. Two million dollars will not make you twice as happy as one million. And in fact, the damage to your happiness if you lost your life savings would probably outweigh your joy at becoming a millionaire of any kind. Thus, you are quite right not to spend all your money on lottery tickets, even in a rollover week.

"The concept of utility explains the third error in your declaration of a few minutes ago. You claimed that you should *never* gamble if your expected return is less than the sum invested. Tell me, Watson, do you carry physician's liability insurance?"

"Of course I do, Holmes: it would be folly not to. Any doctor can make a mistake."

"Yet most doctors never face a liability claim, and I am sure a conscientious man like yourself is even less likely to incur one than the average doctor. The medical insurance union would go broke if the total of payouts exceeded the total of receipts, so on average, as with any kind of insurance, your total life payments will probably exceed your total benefit from the fund. But you are nevertheless wise to have insurance, for by paying each year a sum that does not cause you any hardship, you protect yourself from the potential misery of the poorhouse, and you get peace of mind to boot."

He drew a deep breath. "So you see, Watson, that I could not in all honesty give the Marquis the blanket assurance you wished, that gambling should always be avoided."

"Well, at least you could have told him that gambling on the roulette wheel is never sensible, under any circumstances."

Holmes grinned mischievously. "I can find an exception even to that, Watson. Let us suppose that you are on the island of Krakatoa, and you know that the volcano is about to explode. Unfortunately, the captain of the last ship in the harbor is demanding an exorbitant eighty pounds for passage, and you have but seventy in your pocket. However, there is a dockside casino with a roulette wheel. What would you do?"

"I should go in and place my seventy pounds on red, for at least a 50 percent chance of survival," I admitted.

"You would do better to take the Marquis's advice and use his double-your-money gambling system. Bet ten pounds on the red. If you win, you have just enough for passage. If you lose, bet twenty, and if you lose again, bet your last forty. That way your chance of survival rises to seven in eight."

Holmes's expression softened. "But I admit, Watson, that this is rather a contrived example. In general, your intuition that commercial gambling is best avoided is correct: the house percentage is the only sure winner. I will write to the Marquis tomorrow, and make the point as bluntly as I know how."

3

The Case of
the Surprise Heir

I TOOK MY SEAT AT the breakfast table quietly. Sherlock Holmes
was entirely concealed behind one of the broadsheet newspapers, and I knew better than to interrupt his concentration.
Mrs. Hudson had outdone herself, and for a time deviled eggs
and bacon occupied my full attention. But then my gaze was
drawn to the remaining papers on the table, with the strange-
looking date on the masthead: 1 January 1900. We had
entered the twentieth century. How odd it would be to pro-
nounce that phrase!

I leafed quietly through the pages of the topmost paper.
Shortly, however, I threw it aside with an exclamation of disgust. My friend looked at me quizzically.

"This paper's columnists have made a list of predictions for
the new century. Really, Holmes, I have never read such rub-
bish! On the political front, they predict the decline of the
British Empire, with new powers surpassing us in both the
Western and Eastern hemispheres. They foresee a great new
European war, between France and Germany. Clearly, the
man has been reading Gibbon's work on the fall of the Roman

Empire and thinks he can naively extrapolate from history. Why, Britain has never been so powerful, and Europe never so solidly at peace. I quite understand the human desire to know the future. But of course we never can. Future predictions are always based on the rankest pseudo-science, so why do people believe them so gullibly?"

"I acknowledge these can be no more than informed guesses, Watson, though I have heard Mycroft speculate along similar lines. But I do not believe *all* prediction is pseudo-science. After all, you read the weather forecast, which is sometimes imperfect, yet more often right, and certainly it is based on scientific laws."

"Well, this paper is even more implausible on the subject of future scientific progress. It expects great things of electricity, which is reasonable, if rather obvious. But look at this illustrated page on transportation. Heavier-than-air vehicles taking passengers by the hundreds across the Atlantic, if you please! I am surprised they stop short of predicting a flight to the Moon. All nonsense, Holmes, utter impossibility."

My friend smiled. "Improbable, Watson, maybe. But I have always taught you to distinguish most carefully between the impossible and the merely improbable. Who, a hundred years ago, could have foreseen the telephone, the electric light, the transatlantic telegraph? Do not get too carried away with indignation and ruin your digestion. Nothing you have said contradicts what is physically possible."

I snorted. "There you are wrong, Holmes. The paper is honest enough to quote a well-respected mathematician who says that a flying machine of such a size would have to reach a speed of hundreds of feet per second for its wings to gather enough air to lift it from the ground. Which would require a run of a mile or more along some specially constructed roadway before the thing could even become airborne. Manifestly absurd, you must agree!"

Holmes made no reply, and I surmised he had merely been teasing me. I turned to a more reputable-looking paper, but was distracted by sounds from the staircase. From the creaking, some very heavy person or thing was slowly ascending, accompanied by the rustling of cloth and a metallic jingling. For a moment, I visualized one of those Himalayan pack mules attempting to pay us a visit.

I flung open the door to reveal a scarcely less improbable sight. A woman of enormous girth, of Cockney complexion but dressed in Eastern robes, stood in the hallway. A multitude of heavy metal bangles dangling about each of her wrists explained the jingling. A band of pure white material was tied across her forehead. The strong smell of incense wafted ahead of her.

"I require to speak to Mr. Holmes, immediately," she boomed in an astonishingly deep voice.

"I am afraid you will have to come back later. Mr. Holmes sees no one at this hour, except in dire emergency," I said firmly.

"Emergency!" she said scornfully. "I suppose you refer to petty matters of individual life and death, fortunes won and lost, trivia of but momentary importance before the eyes of Those Who See All. I come to you in an altogether graver context. If Mr. Holmes does not see me now, the consequences will be not local, but universal. The span affected not temporal, but eternal. The followers of a tradition ancient beyond your understanding will be led away from light into darkness, and behind them the whole of humanity. I see you do not recognize me. I am Madam Zelda, High Priestess of the Great Faith!"

She paused to wheeze. One task I must regularly perform for Holmes is to protect him from cranks and crackpots who would otherwise waste his time, and I prepared to dismiss her in no uncertain manner. To my astonishment, my colleague's voice rang out behind me. "Show her in, Watson! I will dress and be with you momentarily."

Seated in our largest armchair—the only one I felt sure would take the strain—Madam Zelda regarded me with, I thought, a touch of pity as we awaited Holmes. Huge dark eyes gazed at me from a lined and wrinkled face. Feeling uncomfortably as though she could read my mind, I was relieved when my friend appeared after only a brief delay. He took a seat and nodded to her to begin.

"Mr. Holmes, I am an old woman. My time is near upon me—no, do not pity me!" She held out a hand theatrically, although my companion had said nothing. "My reward awaits me in the Fields of Azaroth, but a step away from here." She indicated the direction of, I thought, the coal scuttle. "But before I bid farewell with a clear conscience, there is the vexing matter of the succession. The Order whose supreme office I currently hold appoints its guardians by a method more reliable than any vote of cardinals, indeed a method surpassing any human judgment in its infallibility. We are guided, as in all other things, by the stars."

"And how do the stars speak to you?" asked my friend quietly.

She looked at him sharply, but his expression was quite serious. "It is a question of birth date. Using an ancient book, which only the current High Priest or Priestess is permitted to consult, I determine the day of the year on which Zoroaster will be able to part the veil and impart his spirit. The man or woman whose birthday falls closest to that date inherits my power."

"And in temporal terms, that means?" Holmes spread his hands apologetically.

"In temporal terms, I shall make a will to bestow on that person sole title to our temples, which are not churches as you would recognize them but large houses in London and New York, and also a fund of several thousand pounds to be administered on behalf of the Order. I understand your implication, Mr. Holmes: it would be a tempting reward, for one lacking the Faith."

I frowned in some bafflement. "But there are some hundred thousand babies born every day upon the Earth—several thousand in England alone! How can you possibly decide which is the correct one?"

Madam Zelda looked at me coldly. "It would be more than that. It is a question of being born on a given date. The year is irrelevant. By your reckoning, one three-hundred and sixty-fifth part of the world's population would qualify—that is, several million persons. But of course I am not referring to lay people. The indication applies only to true descendants of the First One, who read the stars aright and fled Atlantis when his fellow priests remained behind to meet their doom."

Holmes nodded. "And how do you recognize such persons?"

"Very easily, nowadays. They can only be linear descendants of my great-grandfather, who revived the faith when it was almost extinct."

Or invented it, more likely, I thought to myself but did not say aloud.

"There were sixty-one eligible persons known to me, all but one living in this country. The task appeared straightforward. I divined the date and prepared to anoint my successor. But then an unexpected problem arose. The one who is overseas lives in Canada. He was thought to be the sole surviving descendant of the youngest son of my great-grandfather, who emigrated there. He had long been out of touch, and ironically, I went to some trouble to locate him. For he claims that, far from his being the only one, there are another fifty-nine of his branch of the family living in various parts of the Americas. He sent me a list of birth dates. But I have no way of determining whether these people actually exist. Most impertinently, my cousin will not provide a list of the associated names and addresses. In his letter, he goes so far as to imply that this information might enable me to perform some underhanded machination. He says that if I can tell from this list that one of the American relatives is the Chosen One, I am to tell him the birth date, and he will then take responsibility for contacting my heir."

Our visitor rolled her eyes in consternation. "My cousin's ancestor left for Canada precisely because he mocked my great-grandfather's faith. I suspect, Mr. Holmes, that my cousin is in truth the sole survivor. I have never even met him. If I give the succession to one of his claimed relations, he or one of his accomplices will probably journey here with a forged birth certificate and inherit all the Temple's worldly goods. On the other hand, if I ignore my cousin's claim, I risk overlooking Him or Her whom the stars truly intend to be my successor. Either way, I shall have betrayed my trust!"

Madam Zelda looked at my friend with such obvious anguish that I could not restrain a pang of pity. But he shook his head. "Madam, I see no way to check the claim without myself traveling to the Americas. I recognize the great importance the matter has to you, but I simply cannot undertake—"

She interrupted him, shaking her head vigorously. "There may well be a much easier way. Here are the two lists of birth dates, from the English and American sides of the family." She handed him a small notebook.

My colleague looked at it. "You have not even identified which group is which," he said.

"No," she replied. "You see, I must have confirmation of my guess from you. Very simply, one of these lists looks to me suspicious *in its very nature*. There is something about it— quite apart from its provenance—that is simply not quite right. If you agree with me, it will restore my peace of mind at the least, and save humanity from a dark fate at the most."

Upon hearing these last words, I suddenly recalled how absurd the whole business was. But Sherlock Holmes rose from his chair with an air of utmost gravity. "I shall give it my fullest and most immediate attention. Pray come here at this time tomorrow, and I will hope to have an answer for you."

He escorted her to the door. On his return, he picked up the notebook she had left and inspected it at some length. I looked at him in bafflement. "Really, Holmes, have you gone senile? I feel sorry for the poor woman, but there are matters of genuine importance that await your attention."

"Greater importance than this matter? I think the lady was quite sincere."

"But surely you do not believe in this astral guidance non-sense. Power vested in the stars, forsooth!"

"Of course not, Watson. But there are powers I not only believe in but actively fear, and one such is the insidious power of cults. A growing phenomenon in our midst, both here and in other countries, are those folk who feel a need to put all their trust in a small minority faith under the control of a charismatic leader. Usually they are lost, lonely people, adrift in our society."

"Well, Holmes, who is to say that their faith is a bad thing? It may be misguided, but if it brings them together in a community, banishes loneliness, and provides mutual support, good may nevertheless come of it."

"It may indeed, so long as the leader is sincere, or at least not malevolent. But all too often the cult is either set up by, or soon taken over by, an individual who does not share the supposed faith at all—someone who merely sees an opportunity for ruthless exploitation, be it financial, carnal, or in other ways. The sheep put their trust in a wolf, and the result is disastrous. This is what our recent visitor rightly fears. Fortunately, I shall soon be able to set her mind at rest."

He passed me the notebook Madam Zelda had given him. "It is a problem in pure logic, Watson. Of these two lists of birth dates, one describes real people, and the dates will be genuinely random. The other has probably been invented in the mind of an unscrupulous man. Fortunately, the human brain is a very bad generator of random numbers. Watson, be so good as to pick a number at random, between one and ten."

"Seven."

"Well, illustrated, my friend! Nearly half of all people who are asked that question choose that same number: seven. A

LIST A

Aries	Leo	Sagittarius
March 21	July 28	November 24
March 27	August 2	November 29
April 2	August 7	December 12
April 9	August 13	December 19
April 14	August 22	
		Capricorn
Taurus	Virgo	December 28
April 21	August 25	January 3
April 28	August 31	January 6
May 1	September 1	January 13
May 9	September 8	January 20
May 19	September 18	
		Aquarius
Gemini	Libra	January 22
May 22	September 24	January 27
May 30	September 30	February 4
June 4	October 1	February 14
June 6	October 5	
June 10	October 16	Pisces
June 20		February 20
	Scorpio	February 27
Cancer	October 28	March 5
June 25	October 31	March 11
July 4	November 5	March 16
July 14	November 11	March 20
July 16	November 17	
July 23		

LIST B

Aries	Leo	Sagittarius
March 21	August 8	November 29
April 3		December 14
April 11	*Virgo*	December 19
	September 3	
Taurus	September 21	*Capricorn*
April 21	September 22	December 26
April 21		December 31
April 23	*Libra*	January 7
May 3	October 13	January 9
May 12	October 18	January 9
May 12		January 14
May 21	*Scorpio*	
	October 28	*Aquarius*
Gemini	October 29	January 27
May 31	October 30	February 5
May 31	November 10	February 5
June 1	November 14	February 16
June 3	November 15	February 16
June 5	November 16	
June 9	November 20	*Pisces*
June 18		February 23
		February 24
Cancer		March 1
June 22		March 7
July 13		March 9
July 14		March 11
July 17		March 11
		March 12
		March 12
		March 13
		March 19

fact well known to amateur conjurers and professional gamblers, among others. Of course, a more sophisticated person who knows that fact will consciously try to avoid including the digit seven too often, in generating a pseudo-random sequence of numbers. But in fact there are several kinds of patterns the human mind unconsciously veers toward in attempting such a task. A mathematician who knows this will strive to avoid all those patterns—but in so doing may merely generate patterns of another kind! If this Canadian cousin were a mathematician, he would have picked these birth dates using some formal random procedure, say by throwing dice, or with the aid of a table of random numbers, which can be purchased from a specialist bookstore. Fortunately, it is obvious at a glance that he was no such expert. Tell me, Watson, do you see any fundamental difference between these lists? To my eye, there is a blunder that gives the game away immediately. I only asked Madam Zelda to return tomorrow because advice carries more weight if it seems to arise from lengthy contemplation."

I stared at the lists, determined that for once I should not let Holmes get the advantage of me. "There is one obvious difference. In list A, the birth dates are split roughly equally between the signs. In list B, there are large variations: eleven Pisces, for example, to only one Leo."

"And you conclude?"

"Obviously, that list B is, to say the least, suspicious-looking."

Holmes sighed. "Let us leave that one for a moment, Watson. Look for something more specific."

I took more time to read through the individual dates, and presently I did indeed see something definitive. "I have it now, Holmes. In list B, there are no fewer than seven pairs of persons who each share the same birthday. In list A, there are none."

"Quite so. Now—and try to be a little more careful this time, Watson—which of these situations is more likely to have occurred by chance?"

"I shall be systematic, Holmes. If we take two people at random, the chance that they share the same birthday is obviously 1 in 365. You do not mind if I ignore leap years?"

"By all means do so, Watson; it will make no practical difference."

"On the other hand, with a group of 366 people, a coincidence is not just likely but certain. For even if the first 365 all have different birthdays, every day of the year is then accounted for, and the 366th must share a birthday. So by interpolation, I conclude that if the chance goes from 1 in 365 with two people, to 365 in 365 with 366, then a group of 184 or so would have just a 1-in-2 chance of containing a pair with a birthday in common. But for a group of only 60, the chance is under 1 in 6. You might get one pair by coincidence, I suppose. But seven pairs? I conclude that list B is a fraud, and a careless one at that."

Holmes sighed again. "Let us take this one step at a time, Watson. Suppose you are alone in your consulting room. One patient enters. What is the chance that he shares your birthday?"

"One in 365, of course."

"Very good, suppose that your birthdays are indeed different. The patient's wife enters. What is the chance she shares a birthday with either the patient or yourself?"

"This time it is now 2 in 365. Of course I see now, Holmes. The patient's son enters next, and the chance is 3 in 365 that he shares a birthday with someone already there. The daughter follows: 4 in 365. The probability rises almost as the square of the number of people in the room!"

"Roughly correct, Watson. Now to check your understanding. How many people must there be for the chance of a pair of birthdays to exceed one-half—that is, to be more likely than not?"

I scribbled on the back of a prescription pad: 1/365 + 2/365 + 3/365 + . . .

"It is just 20, which will give a 0.52, or 52 percent, chance of a birthday pair."

Holmes frowned and took the pad from me. He looked at it disapprovingly. "Let us extrapolate in accordance with your logic, Watson. When the 27th person enters, the chance of an overlap, according to you, will be 96 percent. The 28th takes the chance to 103 percent. Better than certainty! Now, Watson, can that really be correct?"

"Of course not. It is certainly *possible* for a group of 28 to have no birthday in common. And I realize that a probability of greater than unity—greater than 100 percent—is a mathematical absurdity in any case. But where have I gone wrong?"

"In a most basic way. The correct way to combine probabilities is not to add them but to multiply them. You know that intuitively. For example, if I toss this coin three times" (Holmes picked up an Egyptian coin from the side table; I recalled that its possession had been the clinching factor in convicting Harris, the poisoner), "then the chance that it will fall heads each time is 1/2 times 1/2 times 1/2, or 1 in 8."

"I know that, Holmes, but I do not see how to apply it in this case. 1/365 times 2/365 times 3/365 and so on gives a ridiculously low figure. That cannot be right."

"The way to do this calculation, Watson, is to turn it around. Two men are in a room. The probability that they do *not* share a birthday is 364/365. A third enters. To obtain the probability that there is still no birthday in common, we multiply by 363/365, for there is a probability of 2 in 365 of that coincidence. For the next man, we must multiply by 362/365, and so on. When the 366th enters, we multiply by 0/365, and the probability that all have escaped the fate of a common birthday falls to zero, as it must. The cumulative multiplications are more easily performed with the aid of a calculating machine, but I can tell you that it is actually when the 23rd person enters the room that the chance of a common birthday rises from about 49 percent to 52 percent. By the time the 60th person enters, the chance of no birthday overlaps is less than 1 percent—roughly 1 in 170, in fact."

"So it is after all list A that is false!"

"Almost certainly, Watson. Seven shared birthdays, as in list B, is much more like the number to be expected. In fact, even without the absence of shared birthdays, you should have been able to tell at a glance that list A was highly suspect. The birthdays were split too evenly among the star signs—and indeed too evenly apart at every scale: there were far fewer birthdays on consecutive days than would be expected, for example. Remember that randomness does not tend to produce uniformity. Rather, data that is *too* average-looking tends to have been cooked! Never trust a man who tells you he tossed a coin a thousand times and got precisely five hundred heads."

I strolled over to the window and looked out. A fresh blanket of snow covered the street to a depth of a foot or so, yet the surface was beautifully flat. It occurred to me that there was a rather neat counter-example here. "But look at the snow, Holmes! Each flake falls in a random place, yet the end result is a carpet of perfectly even thickness. Surely that is an example of cumulative randomness bringing about uniformity?"

Holmes shook his head. "As often occurs, Watson, you are so profoundly mistaken that it is almost poetic. If the flakes really accumulated in independent little vertical piles where they fell, then although the *relative* heights of the piles would become more nearly equal as they grew taller, so that they would look more similar from a distance, this would be despite the fact that the *absolute* differences would be getting slowly larger. The surface of a deep snowdrift would be incredibly jagged—like the stalagmites on the floor of an impenetrable cave, only more so—a surface of needles, with no flat place anywhere. Not a bad simile, Watson: real randomness is a sharp and spiky business, which will cut the unwary as surely as sharp rocks rip apart the boots and hands of the ill-equipped cave explorer. We are unaccustomed to such roughness because processes human and artificial so often give nonrandom pattern to the world we encounter, and uniformity is a simple pattern to generate, and therefore common. The flatness of a real snowdrift, for example, arises because gravity, assisted by the lubrication of the air, causes each flake to tumble to the lowest position in the vicinity of its landing point."

Holmes raised a long finger. "Never mistake uniformity for the product of randomness. Take another example. When a sower sets a field of wheat, the seeds fall in a sense randomly. But the skill of the sower is in fact to give an even covering on the overall scale, and kernels that happen to fall adjacent compete with one another, because a given square inch of soil can provide nutrients and sunlight only for so much growth. The uniform density of the wheat when the field is about to be harvested does *not* arise by chance. But you are not alone in your error: mistaking a uniform distribution for a random one is a common blunder. Indeed, it is worthy of being tagged as the third great human fallacy in misunderstanding the Laws of Chance! You had better start making a list. It is as ever most instructive to talk to you, Watson."

"Do *you* ever read your horoscope, Holmes?" I asked rather mischievously as we sat at the breakfast table the next day.

He regarded me good-naturedly. "Certainly, Watson. It is a harmless amusement, when there is no sufficiently interesting crime reported to give me my money's worth out of the rest of the paper. One can always compare one's expectations with the astrologer's predictions, and perhaps stimulate a thought or two about how to handle the day's business—quite irrespective of whether the predictions are right or wrong, of course."

"Do you not find that sometimes you remember a prediction that has been quite remarkably accurate?"

"Of course, Watson. The key words in your statement are *remember* and *remarkably*. Predictions that turn out to bear no relation to reality are quickly forgotten, nor do you bother to remark on them to your friends. So those predictions that are, quite by chance, accurate are disproportionately likely to be both recalled and mentioned to others."

"You are certain, then, that the stars have no influence at all upon human affairs?"

"None whatsoever."

I held up the morning paper triumphantly. "Then I have news for you, Holmes. You believe in the laws of probability and statistics, do you not? There is an article here by a highly respected mathematician who has uncovered a number of correlations between a person's astrological sign and his or her career, at an incidence he says is quite inexplicable by chance alone! For example, he says professional football players are likely to be Librans or Scorpios. And medical men Taureans or Geminis. Stolid and practical Taureans like myself, or else mischievous and playful Geminis. How well I remember that contrast among my fellow medical students, some of whom were the most irresponsible practical jokers! How can you disbelieve these scientifically proven findings, Holmes?"

"I do not disbelieve them."

"But you just said astrology was all bunk!"

"And so it is, Watson. But just because you measure a correlation between two things—even a perfectly valid and consistent relation—it does not prove that one *causes* the other. In this instance, observe first that the time of year when you are born can affect your life in many plausible ways. For example, the school year begins in September. In a new class of five-year-olds, those who are just turned five, born in August, must compete with their seniors born the previous September, with a grand old age of five years and eleven months. Surely that will make some difference to academic attainment."

Holmes looked up from his breakfast and regarded me sternly. "That is a man-made difference. But as a medical man, you will be well aware that the average birth weight of newborns varies with the time of year. Doubtless factors such as the temperature and the seasonal availability of health-giving foods such as fruit and vegetables have a significant effect on the baby's development within the womb. Infant diseases that can cause later incapacity, major or minor, are also highly seasonal in their incidence. A third possible factor is conception. From your hospital training, do you not remember the flood of illegitimate births at the end of September, the result of merry-making the previous Christmas? It is an easy prediction that some of those poor tots will get less devoted care than more responsible parents would provide.

"Overall, there is every reason that to accept that a man's career should correlate to some extent with his birth date. And of course his star sign also correlates perfectly with that birth date. Thus, a spurious cause-and-effect relationship may appear to operate."

"You have convinced me, Holmes. It is obviously impossible to distinguish between astrological claims and the mundane effect of the seasons."

"Not quite, Watson. You could, for example, study children born and raised in the Southern hemisphere, where Leos are born in midwinter rather than midsummer. Or wait a few thousand years, and the precession of the Earth's axis will gradually change the match between the seasons and the astrological houses. But I feel no need to undertake such heroic investigations. Furthermore, I think we must now shelve this subject, for I hear our client's tread upon the stairs, and I am sure it would be both futile and unkind to debate the matter with her."

He greeted Madam Zelda with dignity, and explained his reasoning to her. To my relief, she nodded at once. "You are right: it is the page labeled A that lists the supposed American birth dates. I had not performed your calculations, but my instinct told me something was amiss with it. Nevertheless, I am relieved to have confirmation."

She looked at us with disconcertingly penetrating eyes. "Of course, I never doubted you would succeed in helping me. A perfect astrological pairing: you, Mr. Holmes, an Aquarian, brilliant, inventive, but unorthodox; and yourself, Doctor, a Taurean, stolid, perhaps even a little slow, but utterly dependable."

Holmes showed her out. On his return, I confronted him open-mouthed. "Now, how could she have known that, Holmes?"

"Known that we made an effective team? I should have thought that that was self-evident, Watson."

"I mean, how could she have known our correct star signs? I am sure she was not yet in earshot when I said I was a Taurean, and I never mentioned your birth date at all. I should think that few know it: I could barely remember it myself."

Holmes smiled at me. "In the newly developed lands of the Americas, where the populace instinctively distrusts authority, there are few central records. It is hard to find out a man's birth date or even to confirm it if he tells it to you. But to find the dates of birth, marriage, and ultimately death of any Englishman, you need only take a trip down to Somerset House and pay your shilling to consult the registers. A quite understandable precaution for one who believes sincerely in astrology and was about to place her trust in two gentlemen she had never met. Misguided she may be, but our Madam Zelda is nevertheless a woman of formidable determination, and I will follow her future career with great interest."

As I prepared to embark on my rounds shortly afterward, I could not resist an attempt to have the last word, after all.

"I am sure you are wrong to discount all psychic claims, Holmes," I said as I donned my boots. "For example, several of my patients still felt persistent pain months after some minor injury, such as a wrist fracture or back strain, had completely healed. They eventually went to a faith healer who practices in my district and then found themselves cured within days. Surely you cannot attribute that to mere coincidence. And I do not believe the pain was imaginary: most of them were conspicuously common-sense, practical people, certainly not the hysterical type."

"I quite believe you, Watson, and I would not ascribe it to coincidence either. Try this hypothesis: a person with an injury learns to hold the affected limb at an unusual angle to minimize the pain. This works to an extent, but the repetitive muscle strain soon causes some pain of its own. Even after the original injury heals, the muscle pain persists indefinitely in a vicious cycle, reinforced by the very action the patient is taking to try to relieve it."

I listened attentively as Holmes went on. "Say such a patient eventually goes to a faith healer, who tells her convincingly that the pain will disappear. For the first time in months, she allows her arm muscles to fully relax, and lo!—in a few days the pain, which was a very real pain with a physical cause, is gone, never to return. Indeed, there is quite a lot of evidence, Watson, that making any convincing positive prediction can actually help someone to do better. Tell a man he is stupid, and he fails an exam; tell him is a genius, and he passes it."

Holmes picked up the morning paper. "And that is a good justification for reading the horoscope predictions in the tabloid papers. Their authors are wise enough to know that their readers prefer optimistic forecasts to bleak ones, and they nearly always oblige. I dare say many a young lad has finally plucked up the nerve to propose successfully, simply because his daily paper assured him that his love life was well starred, and so made the prophecy self-fulfilling."

He looked at the paper he was holding and raised his eyebrows. "But there are always exceptions. Watson, I must advise you to be most careful today."

"Careful? Why will caution be necessary?"

He waved the paper at me. "Because your stars here tell me you will be in distinct danger."

"But danger of what?"

"Danger of believing predictions too gullibly, Watson!"

4

The Case of the
Ancient Mariner

THE PAVEMENT WAS BUSTLING WITH New Year bargain hunters, so
I did not see the men converge upon me until the last
moment. The one ahead boldly caught my eye and nodded as
though he knew me. I hesitated momentarily, for when you
are a doctor, you have many patients whose feelings are hurt
if you do not recognize them; they assume you will remember. In that second, I became aware of other men to my left
and right: I was boxed in. The man ahead pulled a bulky
object from his coat.

Fortunately, I was carrying my life protector rather than my
usual walking stick, for there had been a recent upsurge in
violent street robberies in London, so well publicized in the
papers that you would think the crime was a newly invented
one, though I am sure it is as old as humanity! I took a firm
grip on it: the simple precaution of carrying a stick with a
lead weight in the head has saved my life at least twice. But
a moment later I felt singularly foolish, for the man was
holding not a weapon but a small trestle table. He set it
down on the pavement and dealt three battered playing
cards with a flourish.

"Spot the Lady! Spot the Lady, sir, and win twice your money. A game of pure chance, fair as the young lady there" (a passing shop girl moved on with an indignant sniff), "yet with a little skill as well. Come on sir, double your money: it is as simple as it looks."

I should have disapproved, but somehow I felt more amused than annoyed at the man's importunity. Not since I was a naive young medical student have I fallen for Spot-the-Lady, and then only once. This simplest of all con games has been played on the streets of London for a century, and doubtless will continue for another. Lest any of my readers not have encountered it, the dealer uses only three cards, one of which is the Queen of Hearts; the other two are black cards, clubs or spades. The dealer places the three cards face down, the mark chooses a card, and if it is the Queen he wins double his bet.

There are, however, two twists to the game that remove the element of chance. First, the dealer pretends to be so incompetent that the player actually glimpses the Queen as she is put down and so invariably wins the first bet or two. Emboldened to place a much larger bet next time around, the player finds the dealer less careless—and in fact the player loses no matter which card he picks, for another card has by then been substituted for the Queen.

The two accomplices stood facing away from me. Their role was not to corral the mark, but to keep a lookout for approaching policemen; the game is, of course, illegal. There was nothing to stop me from walking away, and in a moment I should have done so. But then I became aware that the pitch had not been directed at me alone: a prosperously dressed man to my left was eyeing the game with interest.

It was then that a rather cunning idea occurred to me. As a penniless medical student, I had lost the princely sum of two shillings at this game. Remembering that the dealer always lets the first bet win, I felt in my pocket for a half-crown and placed it firmly on the table while my neighbor was still fumbling in his pocket. "I will bet!" I said loudly.

The man dealt. "Round and round and round she goes, where she stops, nobody knows," he called loudly. But just as I expected, he simulated great clumsiness, and the Queen was momentarily visible as she was placed. I picked her and was handed a fresh half-crown. It had taken ten years, but I was ahead of the game!

"Play again, sir?" I shook my head, but before I could warn him, my neighbor was waving a fresh-printed five-pound note.

"I will take this turn," he declared in a strong American accent.

The dealer nodded and made to start again, but at that moment there was a shout of "Rozzer!" from one of the sidemen. In the twinkling of an eye, the dealer picked up the table and stuffed it under his coat, and the three men melted away into the crowd. A moment later I glimpsed an approaching policeman still a good twenty yards away; he had obviously noticed nothing. There was little point in trying to report the matter, and I resumed my walk home, rather satisfied with the affair.

As I ascended the stairs to our rooms, I could tell we had a visitor: Inspector Lestrade's voice sounded angrily through the door. I entered to see him pacing up and down with a face like thunder. Sherlock Holmes, by contrast, appeared in high good humor. It occurred to me that a diversion might be in order.

"Happy New Year, Lestrade. I have just witnessed something that might interest you," I said brightly.

The Inspector barely nodded to me before turning back to Holmes. "Thirteen heads short!" he bellowed. "And yet the magistrates will take no action without 'proof,' they tell me. Thirteen heads missing is not proof enough for them, apparently. It is no wonder that crime still runs rampant, with the police so shackled by the caution of the courts."

Holmes forced his face into a manifestly insincere expression of sympathy. "I will do what I can for you, Lestrade," he said solemnly. "In the case of Davies, I am hopeful I may come up with something. But in the case of the missing heads, I am not optimistic. A month's wages, you say? Still, perhaps you will find some way to make the missus see the matter in a sympathetic light."

"It is easy to see you are not a married man, Mr. Holmes, if I may say so!"

"True, but I would nevertheless suggest a little abstinence. Cut out the beer and the baccy, put the money saved aside, and at least it will soften the blow when it comes."

Lestrade took his leave, still frowning darkly. But as soon as we heard the front door close behind him, my friend threw his head back and burst into laughter.

"What was all that about, Holmes? Surely thirteen missing heads are no laughing matter. I hope we are not back to the days of Burke and Hare, robbing graves to provide cadavers for medical students."

Holmes shook his head, still chuckling. "It is nothing so sinister, Watson. Lestrade is concerned about his new campaign to stamp out illegal street gambling."

"Why, what a pity. I have just had an encounter that would have interested him, had he let me speak. I have become the first man to win his money back from a Spot-the-Lady merchant!" I recounted my experience of a few minutes before.

"It is just as well you did not have a chance to tell him, Watson. Given the mood he is in, he would have been quite capable of running *you* in for illegal gaming. The matter has become quite personal to him. He recently went with a colleague to a pub in Bow Street, right next to police headquarters. There they saw a man who was encouraging drinkers to bet on the fall of a coin. Lestrade was convinced that far from the sequence of heads and tails being random, some sleight-of-hand was going on to force each result, so that some confederate or confederates in the crowd about him were systematically winning at the expense of the rest. Lestrade then bet his colleague a month's wages that he could stamp out illegal gambling in this police area. He is having little success. Unfortunately, he cannot see the funny side: he is about to lose a bet he made—strictly against Queen's Regulations, which forbid gambling of any kind in the police force—because he cannot put a stop to illegal gambling!"

"I am surprised that a zealous officer like Lestrade can make no headway on the problem, though," I said.

"As you heard, it is not that he is unable to find the perpetrators. His difficulty is that he cannot prove fraud to his superiors' satisfaction. For example, he went back to observe the suspect coin tosser, determined to follow him until he had logged the result of one hundred tosses. He counted thirty-seven heads to sixty-three tails. Because most people tend to bet on heads, an excess of tails is just what you would expect if the game was rigged."

"Ah, but you would not expect the division to be exactly fifty heads to fifty tails," I said. "I understand now that you cannot expect things to balance out so precisely. I suppose the question is just how likely is it to get such an imbalance by chance. I would have no idea how to set about calculating it."

"And nor does Lestrade, Watson. That is why he came to me. The curious thing is that he also asked my advice on a second case—a more serious one, yet easier to see the answer to. And the good Inspector does not realize that the two cases hinge on exactly the same point!

"Do you remember that drunk we saw stagger along Baker Street on Christmas Eve? I have just been told his name: Davies. It seems he was a midshipman on HMS *Illustrious,* then moored at Greenwich. He was trying to make his way back to the ship when we saw him."

"That would have been quite some walk," I commented.

"He was not intending to walk the whole way. He had rented a small dinghy and rowed up to Fisherman's Wharf, a derelict dock not far from here, assisted by the incoming tide. He obviously planned that the reverse tide would take him back before midnight, when his leave was due to end. But in fact he never returned."

"Well, considering the state he was in when we saw him, anything could have happened to him."

"And something did. The last part of his walk to the dinghy was particularly tricky. Fisherman's Wharf is a hundred paces long and was in pitch darkness. It is also wide, and at the end there is a sheer drop into the water except for a boardwalk that parallels the end at the central part only, extending twelve paces on either side of the center."

"I do not see how he could have found his way at all," I said.

"Very easily, if sober. The wharf faces due north. At the time he reached it, the Pole Star would have been clearly visible. All he had to do was walk toward it, and he could jump from the end onto the boardwalk, where his dinghy was moored at the center.

"But as we know, he was drunk, and consequently was walking a zigzag rather than a straight line. In fact, from both our own observations and those of subsequent witnesses, he was to all intents walking a perfect mathematical Drunkard's Walk, with no sign of pattern to his veerings, nor any bias to either left or right. He took an intentional step forward, followed by an involuntary one sideways, in a perfectly random fashion a sober man would be hard put to emulate. Thus, although he set out to walk north along the center of the wharf, by the time he reached the end, he had in fact strayed thirteen paces to the right."

"An unlucky number," I commented.

"He was certainly unlucky. The night was cold enough that a thin sheet of ice had formed on the water. Peering down, he saw a solid surface he must have thought was the boardwalk. He jumped down. His keys and other items from his pockets were dredged from the river bed yesterday. The tide must have carried the body out to sea. It probably will never be found."

"Poor fellow. But it sounds like a straightforward enough accident. Why did Lestrade feel the need to consult you?"

"Now we come to the unusual point. Davies had taken out a life insurance policy in the sum of one hundred guineas just two days before. He has no family of his own, and the sole beneficiary was his sister. It is particularly curious that he should have taken out such an expensive policy, because he is known to have had gambling debts. The insurance company believes it might have been suicide—you recall that there were a number of poor wretches who flung themselves into the Thames from various bridges that cold New Year's Eve—and refuses to pay out.

"I believe Lestrade feels sorry for the bereaved sister: the Inspector is less flinty-hearted than he likes to appear. But the insurance company's argument is that Davies was merely simulating drunkenness, ensuring that there were a large number of witnesses to his inability to walk a straight line, before taking a very deliberate leap. Indeed, it may have been no coincidence that his route took him right past the residence of London's most famous detective!"

A shiver went through me at the idea that the drunk who had amused us that night might, in reality, have been walking cold-bloodedly to his death, aware of our gaze upon him. Then another point struck me.

"I am puzzled, Holmes. You said there was a connection with the quite separate coin-tossing case. Whatever can that be? Did Davies's debts arise from that type of pub gambling?"

Holmes smiled. "No, Watson, the connection is a more subtle one. You remember that the Drunkard's Walk can be viewed as a graph of a series of coin tosses. Suppose the drunk spins a coin at each step and veers left if it is heads, right if it is tails. Then his track records the sequence of tosses.

Of course, the distance he finds himself from the center line depends on the relative number of heads and tails he has thrown so far. If he is on the center line, he must have got precisely as many heads as tails. If he is two paces to the left, he must have thrown two extra heads. And if he is thirteen paces to the right—"

"Then he has thrown thirteen extra tails!" I exclaimed. "So to move thirteen paces to the right, in the course of one hundred steps, is exactly equivalent to throwing sixty-three tails in one hundred coin tosses. The one thing is as unlikely as the other."

"You are very quick today, Watson! Now to keep the picture reasonable, I will draw a map of a six-step Drunkard's Walk, rather than the hundred-step one we are really interested in." On a sheet of paper, he sketched a line representing the dock edge, and below it a trellis of triangular shapes.

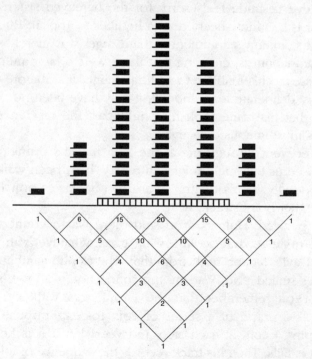

A Walk Along the Pier

"I see: every possible path he can take starts at the base and goes upward on the page, following some route along the diagonals."

"Very good. Now I will write a number at every node—that is, every intersection—of the pattern. Can you tell me what those numbers represent?"

He wrote them in. I puzzled for a while before I was forced to admit bafflement.

"Those numbers indicate the number of different routes he could take to reach that point. For example, there is only one way to reach the top left corner: he must throw six heads in a row, going left each time. But he could reach the point in the middle, the point labeled 6, in six different ways: Head, Head, Head, Tail, Tail, Tail; or Head, Tail, Head, Tail, Head, Tail; and so on."

"There are indeed six routes," I said a while later, having traced them all with my finger.

"I am glad to have your confirmation, Watson, but there is an easier way to calculate the numbers. Simply observe that any point can be reached from a maximum of two others: the two points immediately below it and to the left or right. Starting from the base of the graph, add those two numbers to fill in the next intersection above. The sailor is more likely to reach a point near the center of the wharf than a point near its edge, simply because there are so many more routes to the central part. Can you tell me what his chance of survival is?"

I counted. "Fifty routes terminate at the boardwalk, and only fourteen at the water," I said. "His chance of surviving is about 78 percent. Rather better than I would have thought." I considered further. "I suppose to work out the Davies case, we just need to draw a trellis one hundred steps high," I said. "It will take some time, but I am game. Have we a very large sheet of paper somewhere?"

Holmes smiled. "There is an easier way, which may be familiar from your school days," he said. "Suppose that at the end of each possible walk, the sailor throws down his cap. Each point at the north end of the wharf will acquire a pile of caps." He sketched them in. "Now, does the shape the piles make remind you of something?"

I pondered, and an image from my history textbook came to me. "It looks very much like Napoleon's hat!" I said.

Holmes sighed. "To most people, it is more reminiscent of a bell," he said. He fetched down a well-thumbed book, and opened it to show a smoother version of the curve. "This is a graph of the shape the cap pile would take as the number of coin tosses tended toward infinity: the famous Bell Curve. A very large number of walkers starting from the point X and taking very small steps to reach the horizontal line, on which they then place their caps, would form a pile of the shape shown."

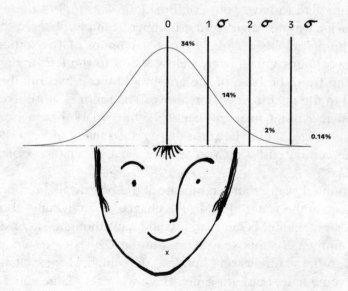

*Napoleon's Hat or The Bell Curve
or The Normal Distribution*

I nodded, but privately I still thought the curve looked more like Napoleon's hat, and I penciled his face beneath to demonstrate the point.

"A significant parameter, Watson, is the average distance that a cap falls from the center line. It is called the standard error or standard deviation, and it is properly indicated by the Greek letter sigma, which looks like this." He wrote in a σ following the numbers aligned above the curve. "There is an easy way to compute this standard deviation if the walk is of reasonable length: it is roughly half the square root of the number of forward steps. For a hundred-step walk, it is therefore five paces.

"Now we have but to look at the areas of the graph. Roughly one-sixth of the time, the number of tails will exceed fifty-five, one standard deviation to the right of the center. Only one-fiftieth of the time will it exceed sixty, two standard deviations. Sixty-three tails or more, as in the pub game, occurs in under half a percent of cases. Of course you might argue that sixty-three or more heads would be equally suspicious, but even taking both possibilities into account, an imbalance as great as Lestrade observed would occur by chance only one time in a hundred. Similarly, a drunk following Davies's system for reaching the boardwalk where his dinghy was moored would stand only a 1 percent chance of ending up in the river."

"It would seem to be good news for Lestrade, but bad news for Davies's sister, then?"

"You are correct. Given that there were already good grounds to suspect both Davies and the coin gambler, I think it is possible to conclude beyond reasonable doubt that both are guilty."

I mused for a while. "I must say, Holmes," I ventured at length, "you were fortunate that some mathematician had devoted himself to working out this graph you had at hand, describing the distribution of large numbers of coin tosses."

My friend smiled. "This graph applies not merely to coin tosses, Watson. Its proper name is the Normal Distribution. Of all the nontrivial graphs known to mathematicians, it is the one with the most universal applicability to the real world. Take any population of nonidentical objects you care to mention—animal, vegetable, or mineral; natural or artificial. Pick any quantifiable characteristic of those objects, subject only to the restriction that the characteristic is at least partly under the control of a fair number of randomly varying influences."

I had no idea what he was getting at. To demonstrate this fact, I sought a deliberately ridiculous example. "The heights of red-haired dairy-maids in Yorkshire," I said.

"Excellent example, Watson. Go to Yorkshire, and measure a reasonable sample of such, say one hundred, sufficient to give you a reliable estimate both of the average height and of the standard deviation. You can then be sure that the whole population of dairy-maids conforms to this curve: for example, that just one in six is more than one standard deviation taller than average. You would even be able to extrapolate from your sample—to know what proportion of the maids were over six feet tall, say—even though you had not actually encountered a single example of such a girl."

I must have looked doubtful.

"The curve arises inevitably, any time that you are summing or combining more than a few variables, Watson. Tell me, have you heard of Charles Booth's new map of London?"

This seemed to me a complete non sequitur, although I remembered reading about it. "As I recall, Booth has been studying poverty in London, and he has produced a color-coded map to highlight the differences in prosperity between different areas. For example, gold denotes a well-off street, and gray indicates what he calls 'vicious and semi-criminal' inhabitants."

"Correct, Watson. Now, a new board game has recently been designed that I strongly suspect, although I have no proof, may have been inspired by Booth's map. Players move tokens clockwise on a single path whose squares denote properties around London. The squares are color-coded from brown, which denotes the Old Kent Road, to a royal purple, which is Park Lane. Players may erect houses on properties they have purchased and, having done so, may charge appropriate rents to other players who subsequently land on the squares representing those properties. It struck me as a promising idea, and I have purchased some shares in the company producing it."

"But if movement is governed only by dice throw, there cannot be much skill to the game. It will never catch on, Holmes. Take my advice, and sell the shares while you still can!"

"Actually, I think I shall hang on to them for a while, Watson. But there are admittedly a number of unresolved questions in the game's design. Players often build houses on the basis of forecasting that some other player may be just about to land on that square. For maximum excitement, therefore, the distance a player moves should be neither completely unpredictable nor completely certain. The predictability is in fact governed by the number of dice used. Be so good as to draw me a bar chart of the outcome of rolling a single die."

It seemed an idiotic task, but I drew the top chart reproduced on the next page. "Every number is of course equally likely, Holmes."

Holmes nodded. "Next, assume you throw two dice and move the distance given by their sum. Draw me a chart of that beneath the first."

I did so. "Now you have a pyramid. It is obvious that you are far more likely to throw, say, a seven than a two. Because there are six different combinations that move you seven squares: 6 with the first dice and 1 with the second, or 5-2, or 4-3, or 3-4, or 2-5, or 1-6. And there is only one combination, 1-1, that moves you two squares. It is a little like the Drunk-

One die

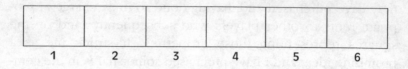

Two dice

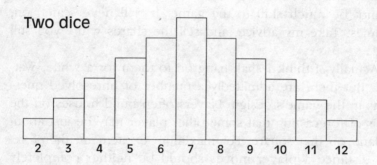

Three dice

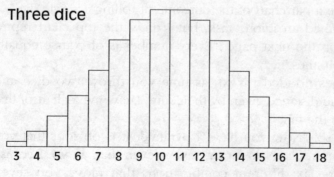

Sums of Dice Throws

ard's Walk, Holmes. You could say that in some sense, there are more paths that will lead the player to a spot seven squares ahead than to the extreme values of two and twelve."

"Indeed there are. But now, what if you are playing a variant of the game in which you throw three dice to move?"

I drew this final chart below the other two. "You are now even less likely to get the extreme values, three or eighteen, and even more likely to score in the middle of the range. Why, confound it, Holmes, I have once again drawn Napoleon's hat!"

"Quite so, Watson. You have indeed chosen a fitting name for the Normal Distribution. Just as Napoleon sought to conquer all the populations he encountered, so the 'Napoleon's hat' curve tends to dominate all random populations encountered in nature. But remember this: Napoleon ultimately failed in his quest—he never ruled *all* of Europe, despite his ambition. And similarly, not *every* imaginable population conforms to the normal distribution, although student mathematicians sometimes fall into the trap of thinking that all must."

"I shall remember, Holmes. I now feel most thoroughly knowledgeable in matters of statistics."

"Good! Then I am happy to leave you to explain the matter to Lestrade."

"Me? But surely you yourself would do it so much better."

"I have an errand to perform, and Lestrade is due to return in an hour. I hope to be back myself not long after that, but I am happy to leave the explanation in your capable hands."

I rose in some alarm. "Unfortunately, Holmes, I have just remembered a task of the utmost urgency, which will take me all afternoon. Hence, although I am honored by your faith in me, I must regrettably decline—"

"What task is that, Watson?"

To my embarrassment, I was quite unable to invent one in time! Nevertheless, when Lestrade arrived, he seemed to follow my explanations with a minimum of difficulty. He was delighted to learn that he could arrest the pub gamester, but less pleased at the news that Davies's sister was unlikely to receive the insurance money. Apparently she was a widow with several children to support. It did seem very harsh that Davies should have thrown his life away to no avail, and looking back, I was a little surprised at Holmes's callousness. At this point, however, my friend returned, accompanied by a white-bearded seaman.

"Good afternoon, gentlemen. May I introduce Captain Darling, formerly of the fisheries and now retired? It turns out that he was taking a stroll on Fisherman's Wharf on the fateful night, and his eyewitness evidence proves beyond all doubt that Davies's death was after all an accident."

"An accident?" I protested. "But what of your carefully worked out probabilities?"

"Never confuse probabilities with certainties, Watson! Sometimes 1 percent chances come off; sometimes things even unlikelier than that happen, as any experienced mariner will tell you. One must always seek to turn probability into certainty wherever possible."

Lestrade looked at Holmes with respect. "I must admit that was quick work, Mr. Holmes, to turn up a witness so fast. And a good thing you did: I was about to make an error, for Dr. Watson's statistics were really quite convincing."

Holmes shrugged. "There was nothing to it, really. I had only to go along the dockside pubs, explaining that a deserving lady would not get her money unless a witness could be found, and sure enough one came forward. Captain Darling will tell us the sad tale. Please step forward, Captain."

I was slightly surprised that he had not offered the captain a chair; such a failure of manners was unlike him. But that was nothing to his next breach of the hospitality a host owes a guest. As Darling shuffled closer, Holmes stepped up to him, seized his bushy white beard in both hands, and pulled violently downward. Darling gave a cry of pain, but a moment

later the entire beard pulled free with a tearing noise, revealing the unshaven face of a much younger man.

"Gentlemen," Holmes cried, "meet Midshipman Davies, of
Her Majesty's Navy! Lestrade, please be so good as to arrest
this man. He can be charged with desertion from HMS *Illustrious* as well as attempted fraud, for he is a good deal more
than twenty-four hours overdue from leave."

The following day, I found myself still pondering that curious
Napoleon's-hat curve. "After all, 1 percent chances do happen—1 percent of the time! Davies could have been innocent," I muttered, mainly to myself. But Holmes looked up
from the book he was reading.

"Indeed he could, Watson. And so I was glad to have
irrefutable confirmation. But we live in a world of uncertainties. When using that bell curve in front of you, statisticians
often consider something reasonably certain if it is 95 percent
probable."

"When the criminal law demands 'proof beyond reasonable
doubt,' is that a reference to this 95 percent level?" I asked.

My colleague shook his head. "The judicial system, both here
and in the United States, has always been unwilling to set 'reasonable doubt' at any precise percentage value. Wisely, in my
opinion. In practice, it seems juries usually require a higher level
of proof to convict a defendant for a more serious offence—say,
murder as opposed to dishonest gambling. We should always
remember that the artificial absoluteness of a Guilty or Not
Guilty verdict usually rests on probability rather than certainty. I
think the Scots are wise to allow also the Not Proven verdict for
cases that lie too close to the dividing line to call."

Holmes sighed. "Judges are aware of the uncertainties, of
course. I suspect they sometimes take it into account in their
sentencing, even though in most contexts they are not supposed to. You will never hear a judge say in so many words,
'The jury considers your guilt certain, but I think it is merely

rather probable, so I am letting you off with a light sentence,'
but I am sure it happens. The Law is no more perfect, Watson,
than any other instrument of man's devising."

I paused to stuff my pipe and consider the dilemma of
uncertainty. "You did not answer my question directly,
Holmes, but from your actions your own opinion would seem
to be that 95 percent certainty is sufficient to convict for mod-
erate offences, whereas 99 percent should be required in the
most serious cases," I said eventually.

"I would not deny it, Watson."

"Does that not mean that murderers are more likely to go
unpunished than petty offenders, though?"

"No. It means that we expect the police to put more effort
into investigating serious cases such as murder, even after they
have found a prime suspect, because it is important to be as
nearly certain as possible before you execute someone."

I considered the ramifications. Eventually another thought
occurred to me. "I suppose that means I can abandon my pipe
and start smoking cigarettes again," I said.

Holmes looked at me with raised eyebrows. "How on earth
do you reach that conclusion, Watson?"

"Well, there is a certain amount of evidence that cigarettes
are bad for your health. But the manufacturers point out that
in the here and now of A.D. 1900 the case is not yet proved—
certainly not at either the 95 percent or the 99 percent level of
certainty. It would be unfair to boycott their products in those
circumstances, until more evidence emerges."

"Really, Watson, at times I wonder at you! We must demand
a high level of proof in criminal trials for reasons that have to
do with human rights, and also because of the dangers of the
process of law being abused—by false allegations, for example. But a cigarette does not have any human rights. You do
not have to wait until it is proved guilty before executing it!

"The level of proof demanded in civil cases, where only
money is at stake, is not 95 percent but 50 percent; they are

decided on a simple balance of probability. And you are perfectly entitled to take still lower probabilities into account in your everyday decisions. If the cost of avoiding a danger is low, and the potential outcome is grave, then a very small risk—even one that cannot yet be accurately quantified—is still worth avoiding. Consider a captain who has a hunch that there may be icebergs in the waters ahead of his ship. He is not 95 percent certain of it, or even 50 percent certain, but he is quite within his rights to reduce speed, double the watch, and offer a tot of rum to the hand who spots the first berg. Better late than sunk. If you must smoke, Watson, take my advice and stick with your pipe. In any case, it really suits your image so much better!"

5

The Case of the
Unmarked Graves

"Go, Watson? Of course you must go! I urge you precisely because your aid is so invaluable. A Watson with fresh eyes and recharged batteries will come back to me twice as useful as one worn down by the August heat."

I must confess to some relief at hearing my friend's words. Sherlock Holmes enjoys London in every season, his fascination at the machinations of its three million inhabitants outweighing any petty discomfort caused by extremes of climate. But I have always found the dog-days of August pretty much intolerable: when the wind blows from the east, the combination of the heat and the stench would make any sensible man seek pleasanter climes. With a competent colleague available to oversee the care of my patients, the invitation from my old college friend Prendergast had been as welcome as it was unexpected. However I knew Holmes was in the midst of one major and two minor investigations, and I had hesitated to leave him in the lurch.

Nor did his reassurances fully assuage the guilt I felt as the hansom rattled toward Paddington. For the fact of the matter was that I had been less than frank with him. Prendergast's invitation was not merely to enjoy the countryside but also to investigate a crime. And the crime was one of mass murder!

I have long felt that familiarity with my friend's methods would make me quite a competent investigator in my own right. But I never have a chance to prove it, for Holmes is undeniably quicker than I: when we work on a case together, he is always at least one step ahead. Of course, ordinarily I would never even consider cutting him out, risking a delay in solving a crime, during which the perpetrator might be free to strike again. But no one could claim time was of the essence in this particular case, for the murders (if such they were) had taken place over a thousand years ago.

At Paddington I tipped the cabby rather lavishly and found my seat on the romantically named Penzance express with plenty of time to spare. I expected the train to fill up somewhat, but when it pulled out I still had a whole carriage to myself. Even when it stopped at Reading, no one boarded except a couple of schoolgirls. As we rattled on into open countryside, I felt my spirits beginning to lift. England is a crowded island, but the West Country is its deep back garden. Even today its pleasant fields are still interspersed between dense forests and moorlands where few venture: exactly the tonic needed when one has been too long in London's crowds.

The train rattled over bridges as it crossed and recrossed the winding path of the Thames, and in no time the dreaming spires of Oxford came into view. The train stopped briefly to permit a shy-looking clergyman to board. I rather expected him to sit within conversation range of me, but instead he took a table opposite the schoolgirls and was soon entertaining them with various card tricks. The bright sunlight and girlish laughter were a soothing combination, and before long I must have drifted off to sleep.

I woke with a start to find the sky much darker and the train gathering speed as it pulled away from a platform. In something of a panic I scrambled to peer through the window lest I had just missed my station. I heard a shy cough, and, looking around, I saw that the schoolgirls were gone but the clergyman was sitting just as I had last seen him.

"This is Bristol we are now leaving. Merville is some hundred miles ahead of us yet," he told me.

"Thank you—but how did you know my destination?"

"The ticket collector told me. That is my stop also."

Having ignored me earlier, the little man now seemed anxious to make conversation. He pointed past my shoulder. "You see that castle?"

I beheld a huge edifice on a hilltop, dominating the valleys on either side. "One could hardly miss it."

"It is not merely imposing but also very old. Some say it was King Arthur's court, although it is not really far enough west for that to be plausible."

I smiled. "I have a friend who would certainly be most disappointed, were that proved to be the case."

He looked at me intently. "Would your friend's name be Prendergast, by any chance?"

"The very same: I am on my way to visit him now. You are also acquainted with him? My name is Watson: Dr. John Watson, at your service."

The little man beamed and half rose so that he could reach to shake my hand.

"I am also on my way to visit him. The Reverend Charles Dodgson; delighted to meet you. But I think I recognize your name: you are the colleague and biographer of the famous detective Sherlock Holmes, and the author of those popular stories of his cases that have met with such public success."

He smiled shyly. "My own writings are confined to rather dry mathematical papers and sermons. But I do have a certain facility for entertaining children, and have sometimes wondered if I could write some lighthearted fantasy for them worth publishing."

"A good children's book is worth its weight in gold," I said warmly.

"There again, I might try my hand at a book of mathematical puzzles and games. Do you think such a thing might sell?"

I could barely repress a shudder. "I am afraid that to me, mathematics for enjoyment sounds rather like a contradiction in terms."

He pursed his lips. "Oh, I think there is a certain fun to be had in questions of logic and probability, and in the surprises and paradoxes that can arise."

"Probability? Well, I am no mathematician, but I think I now understand that subject: the chance of tossing a coin and getting a head is a half, and two successive heads a quarter, and so on. Probability is such an intuitively obvious matter that I doubt you could devise a puzzle that would fool me!" I said.

My companion smiled. "Heads and tails: yes, perhaps such binary choice is rather straightforward. But suppose you have *three* choices, now. Cultures ancient and modern have regarded three as a mystic number, beyond ordinary intuition. Perhaps three-choice puzzles could be quite a different matter."

I shook my head firmly. "I am confident the logic would be no harder," I said.

The Reverend hesitated, and then evidently felt it was time to change the subject. "How is it you come to know Prendergast?" he asked.

"We were in medical school together, but he was always more interested in English language and history; he dropped out halfway through the course to do English at Oxford. That is doubtless where he met you?"

"Indeed, he was at my college, Christ Church. We remember him well. He was one of those undergraduates whose expertise came to surpass that of his tutors in some areas, notably Anglo-Saxon and Nordic runes. And of course he has gone on to make quite a name for himself as an amateur archaeologist. You are possibly familiar with the ambition that drives him?"

I had to smile at the notion anyone could know Prendergast for more than a day without hearing the story.

"Of course—his notion that his family are the descendants and rightful heirs of King Arthur's throne. I am afraid we ribbed him about it at school. He reckons that Prendergast is a corruption of Pendragon, Arthur's family name, and that Merville itself is named after the sorcerer Merlin, an abbreviation of Merlinsville."

The Reverend looked at me reprovingly. "It is certain that his family is one of very few whose title dates back to before the Norman Conquest," he said. "Like the Romans before him, King William found it easier to make peace with the westernmost parts of England than to take his army so far from its heartland. Superstition may have played a part in his decision: the bleak moors and impenetrable woods of the further West Country are frightening to soldiers from other parts, likely to tempt them to desert or mutiny. And of course the infamous pirates and smugglers of Penzance continue to operate beyond the effective reach of London's law and customs officers to this very day. I deem it quite possible that Prendergast's family are descendants of the original kings of the West."

"But surely, as a Church of England man of the cloth, you cannot believe in the stories of Merlin and black magic and so on!" I cried.

"Not literally, and of course a thousand years of oral tradition have embellished the original truth. But let us not argue yet. If Prendergast's letter to you is like that I received, he wants our assistance in interpreting remarkable new archaeological evidence that he has found. It is invariably a mistake to theorize ahead of the evidence."

I started: for a moment he had sounded almost exactly like Holmes!

"His account to me was rather short on detail," I said. "I gather he has found the remains of an ancient graveyard and hopes to prove it is the resting place of his ancient ancestors. I am a little puzzled as to why he waited to summon us, rather than pressing ahead with his dig."

"I should imagine he wants witnesses to the disinterment. Sadly, amateur archaeology has often been marred by deceit and the fabrication of evidence. If a doctor and a priest, the former widely known for his writings, are present at the excavation, it will safeguard him against any suggestion of fraud."

"It is a wise precaution, given his personal interest in the outcome," I agreed. "I gather that as well as authenticating his ancestry, he hopes to dispel lurid stories about how royalty in those days met their end. I am optimistic on that score: if local legend has spun stories of magical deeds out of thin air, tales of human sacrifice are doubtless equally specious. After all, King Arthur's court was a chivalrous Christian one."

The Reverend pursed his lips. "I am afraid I do not put too much trust in that," he said. "Christianity came here in Roman times. But after the Empire receded, its tradition almost died out locally; it survived only by cross-breeding with more ancient pagan beliefs. You have no doubt heard of the Druid priests? The results of such corruption can be very dark. Shakespeare put it well, as ever: 'Lilies that fester smell far worse than weeds.' We must prepare our friend for the worst, Doctor."

We made lighter conversation as the train ran on through gradually thickening woods. But my pleasure in the scenery was quite spoiled. The sinking sun threw ever darker shadows, and soon it was as though the train were running through a tunnel. I felt a superstitious chill run through me, and did not know whether I was relieved or alarmed when at last the guard's cry came: "Merville next stop!"

The train paused only momentarily, and I helped the elderly Reverend with his luggage as we disembarked in haste onto a wooden platform. We were the only ones to alight there, but a man waved to us from a four-wheeler parked just beyond. Seconds later, Prendergast was pumping our hands enthusiastically.

"Glad you could make it, Reverend. You too, Watson. You are already acquainted, I see. It is delightful to see colleagues from both my medical school and Oxford days together."

We boarded the four-wheeler and rattled off into the gathering gloom. Prendergast peered at us both. "You must be hungry from the journey, and dinner is waiting, but there is something I want to show you en route: some stones just a few yards out of our way. May I have your indulgence?"

We both protested that we would be delighted, although my stomach was beginning to rumble. Shortly the horse turned onto a side road but halted just a few yards down. Beyond, there was some indication that the path continued, though it was very overgrown. Beside us was evidence of recent digging. We scrambled down to see four shallow holes. At the base of each was a crude, ancient-looking flagstone from which the earth had been carefully scraped and brushed. Prendergast held the coach lantern down so that we could see how the four flagstones were marked.

The Counting-Stones

"You think these are grave-markers?" I asked.

"No: they are only what are called counting-stones. They are found in the vicinity of graveyards hereabouts. No one knows their precise significance. Each stone always has a rough circle scrawled on one side and anywhere from one to twelve straight lines marked on the other. What the count indicates, no one knows. Fortunately, that is not the point of importance.

"The wood beyond is forbidden territory. There is an ancient injunction that no one may walk there except the Knights of the Round Table themselves. That still has the force of law, and only my father can grant an exception. This he will do only if I can convince him that what lies beyond is a graveyard of our own clan.

"These stones could pretty much prove the fact. For I know from my research that our clan had a peculiar superstition. The circles are always drawn quite crudely and may be left either open like a horseshoe or closed. But our clan, and our clan only, followed an additional rule: the circle must *never* be left open *unless* there are five or more lines on the other side."

"Then surely you have only to dig a little farther and extricate these four?" I said. "If none of the four violates the rule, that would be pretty good evidence."

Prendergast nodded. "My father concedes as much. There is a snag, however. He is so reluctant to permit violation that he will allow only two of the four stones to be extracted. And he insists that I can go further only if I can show that none of the four breaks the rule. He is a very difficult man, gentlemen, a most cunningly stubborn man!"

Dodgson frowned and seemed about to speak, but Prendergast hurried on. "What is particularly tantalizing is that I have found a further stone that, were I permitted to lift it, would prove the matter beyond all doubt."

He led us some fifty yards around the edge of the wood, where our way was interrupted by a hill. Obviously ancient, it nevertheless had the symmetrical look of something placed by design rather than by Nature. Prendergast pointed to three large blocks of stone protruding from the mound. Two were mostly buried with only the ends showing, but the one nearest us was more exposed, and we could see most of its top face.

Prendergast held the lantern above it, and I saw that this block was marked with a pentagram whose five points were clearly visible.

"Now, these are burial stones, and of an interesting kind. Their thickness indicates that each marks the resting place of a married couple. The symbol of each partner is inscribed on one side of the stone: a pentagram for a member of royalty, a simple triangle for a commoner. If my researches are correct, then one of these stones is for a royal couple; one for a pair of commoners; and one for a mixed marriage, commoner to royal."

"So one stone has a pentagram on each side, one a triangle on each side, and one a triangle and a pentagram," said the Reverend thoughtfully. "I suppose in the case of a mixed marriage, the pentagram would go on the upper side?"

"No, the upper symbol should be that of the elder of the couple. We have no way of knowing their ages, so which symbol goes uppermost is perfectly random from our point of view.

"Now, only one of these three graves is really of interest, and that (as you might guess) is the royal-royal combination. It is likely to contain treasures of historical significance; the others will have nothing but bones.

"But my father will not permit any digging into the mound on any pretext. He would permit the excavation of this protruding stone, but only if I can prove that there is a better than even chance it is the purely royal one. Although you see the royal symbol on top, there is obviously a 50 percent chance the lower symbol is that of a commoner. So I am allowed to go no further. It is really very frustrating!"

I felt most sympathetic. But as we walked back to the coach, I was surprised to see Dodgson smiling. He placed a hand on Prendergast's shoulder. "Pray do not fret too much. I have a feeling—a distinct feeling!—that I may be able to persuade your father to think favorably of your chances."

Prendergast only shook his head. "My father is not accustomed to compromise. In the powers he wields, he is in some ways more like a feudal lord than an ordinary country squire. His formal title, by the way, by which you should address him, is the Mage. Under the treaty granted him by William the First, he even has the right to retain his own private army—something normally prohibited by British law."

Dodgson nodded. "Not quite uniquely, I think," he said. "I believe at least one of the Scottish Highland chiefs retains that right, and the rulers of the Channel Islands, especially the Marquis of Sark, continue to exercise near-feudal powers to make and enforce law locally. But I appreciate that we must tread cautiously in any dealings with your father."

By now I was braced for something out of the ordinary, so I was not unduly surprised when we came into the driveway of Prendergast Hall and saw a building that looked more like a modern version of a medieval castle than a country house. There were uniformed soldiers on guard, and I noted that although each carried a ceremonial halberd (an ornate combination of ax and spear), they also wore holstered revolvers. But within, the furnishings were modern and the temperature comfortable. The Mage was nowhere to be seen, and we partook of an informal but generous buffet supper. Prendergast was summoned away during the meal; he returned with a somewhat embarrassed expression.

"I say—it is quite an imposition—but I had completely forgotten that today is the day when my father puts on an entertainment for the youths of the surrounding villages. Like most things we do here, it is a long-established tradition, but there appears to be a complication tonight that I hesitate to leave the servants to handle."

The Reverend and I assured him that we were ready to offer any assistance.

"The problem is that we provide cider for those aged eighteen and over, if they wish it. For younger people, and of course for any older ones who prefer it, we have orange juice. Those invited have all been issued a ticket with their age printed on one side. On the reverse side of the ticket is written a C for cider or an O for orange, entitling the holder to be served the appropriate drink by the bartender. The letter C or O was stamped in indelible ink when the ticket was purchased, for cider drinkers were charged more.

"Now, the ticket sellers were supposed to check that no ticket was stamped C unless the age on the other side was 18 or more. But one ticket seller failed to do this. So we need some responsible person to check each ticket at the door and make sure that no underage tickets marked C are slipping through. Reverend, you are good at dealing with young people, are you not?"

The Reverend shook his head. "I am fond of children, except boys!" he said. "Of course, if the Doctor is willing to do the actual inspection, I will be happy to stand beside him and lend moral authority."

So it was that I found myself standing guard at a side door of the castle, with the Reverend as my companion.

"Poor Prendergast," I remarked to him. "If his father will not even let him turn over the four counting-stones as a first step, I do not see how he will get any further."

"On the contrary, Doctor, if you think hard, I believe the solution will occur to you."

At this point, however, a group of four youths approached. The first showed me the front side of his ticket. The 19 I saw there entitled him to any choice of drink, so I let him through without checking the obverse. The next guest showed the reverse side of his ticket: it was O, so again I let him through without needing to turn the ticket over to check his age. However, the third youth showed a ticket marked C, and I made him turn it over to see that the other side was marked 20: he was permitted the cider. The fourth youth showed an age of only 15, so I made him turn the ticket to verify that the reverse

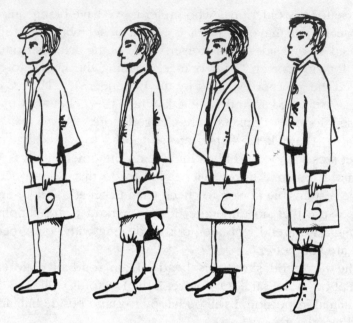

Checking the Queue

was marked O. For some reason, my actions seemed to amuse Dodgson immensely.

"You think I could be handling the job better?" I asked with asperity.

"On the contrary, Doctor, you are a model of efficiency. I assure you, your colleague Sherlock Holmes could have done no better!"

Before I could inquire further, we were engulfed by a wave of party-goers. The stream did not let up until the doors were closed at 10 o'clock, and a footman led us back to the smoking room to enjoy a drink ourselves in quieter surroundings. Prendergast was awaiting us, accompanied by an older, very tall man with a sun-darkened face and piercing eyes. This, at last, was the Mage. He nodded coldly as we were introduced.

"I understand Edward has invited you down to witness his tomfoolery. You are welcome guests, of course, but one thing you will not do, and that is cause me to go back on any edict I have laid down. Once my mind is made up, sirs, it is made up!"

"Then when you say that we must prove that all four marker stones follow the clan's tradition, we must prove that by turning over only two?" said Dodgson politely.

"Indeed," said the Mage with a harsh twist of his lips.

"Why, even Watson here knows how to do that," said Dodgson. I looked at him in astonishment. Dodgson took out a notebook and opened it to a diagram of the stones we had seen earlier.

"The rule we are testing is that any open circle must be matched by five or more lines. The first stone bears an open circle, and we must turn it to count the lines on the back, for if there are fewer than five, the rule would be broken. The second stone might have an open circle on the underside, and if so it would break the rule. So again we must turn it to be sure. But the third stone has a closed circle. That cannot violate the rule no matter how many lines are on the far side. Similarly, the last stone has six lines, and so it will not break the rule whether the circle on the underside is open or closed."

"That is very clever," I said. "But whatever did you mean by saying that I knew the solution?"

"Why, you just solved the very same problem in your role as ticket inspector. The rule there was that a ticket with a C for cider—that is, a broken circle, as opposed to a closed O for orange juice—must have a number of at least 18 on the back. You immediately recognized that if you could see either a number greater than or equal to 18, or an O, on whichever side of the ticket was toward you, it was unnecessary to turn it over. The rule was identical to that for the stones, except that the threshold number was 18 rather than 5. It is rather striking, is it not, Doctor, that when an abstract logical puzzle you could not solve was translated into an identical one that involved human interactions and possible cheating, the solution was obvious to you even without conscious thought!"

The Mage smiled. "You are an ingenious man, Reverend," he said. "You may test the counting-stones tomorrow, as you suggest. But it will get you no further, because I will not permit excavation of the stone protruding from the barrow unless you can demonstrate that it is, more probably than not, the royal one. And the chance of that is clearly only one-half: the top is of the stone is marked royal, but the bottom is equally likely to be royal or common, which is not good enough."

The Reverend merely nodded, quite unintimidated. He drew from his waistcoat pocket three cards and turned them so that we could see that one was green on both sides, one red on both sides, and the remaining one green on one side and red on the other.

"Let us reproduce the problem as a child's game," he said. "So that you are assured there is no cheating, I will number the sides of the cards."

As we watched, he numbered the all-green card's sides 1 and 2, the all-red card's sides 3 and 4, and the remaining card 5 on its red side and 6 on its green side.

"Let us suppose red denotes a royal, equivalent to the pentagram, and green a commoner, equivalent to the triangle. The three cards match the stones in the barrow. I will shuffle the cards concealed in a napkin"—he did so—"and then draw one partly out so that you can see most of the top side, but not the number. Why, I am in luck! It is red for royal, just like the top side of the stone sticking out of the barrow."

Ignoring the Mage's patronizing smile, the Reverend continued. "Now, what is the chance that the obverse is red? There were six sides in all, each in principle equally likely to come up, but we know it is red, so it must be side 3, 4, or 5, with equal probability." He held up his hand and counted on his fingers. "If it is side 3, then the obverse is red: side 4. If it is side 4, then the obverse is red: side 3. If it is side 5, then the obverse is green: side 6. The chances are therefore two in three that the reverse side is red. Therefore, the chance that the stone protruding from the barrow is all royal is not one-half but two-thirds. And by your wisely chosen rule, Mage, we are permitted to excavate."

The Mage's jaw muscles clenched, and his face turned purple. For a moment I thought he might be about to forget himself so far as to strike a man of the cloth. Then he gave a stiff nod and strode from the room without a further word.

Prendergast turned to Dodgson and wrung his hand silently. "Reverend, you are just as much of a wizard of logic as I remembered. Tomorrow we shall start our quest!"

In the morning we set out accompanied by four of the Mage's men carrying shovels and crowbars. The Mage himself followed at a distance. The walk to the site of the counting-stones was shorter than I had expected; it had seemed longer in the dusk of the evening. At Prendergast's orders, the counting-stone with the broken circle was levered over to reveal seven lines on the other side, and the one with three lines was turned over to reveal an unbroken circle. So we had confirmation—or, to be more precise, no actual refutation—of the notion that the stones had been made in accordance with his clan's rules. We moved on to the barrow. I held my breath as the much larger stone with the pentagram was loosened and slid out. Even with two crowbars, it was all the men could do to turn it over, and Prendergast and I bent our shoulders to add our weight to help them. But our efforts were rewarded: the stone crashed over to reveal a second pentagram. The men dug with a will into the barrow, and four feet beneath the stone's original resting place, their shovels struck something that caused them to drop their implements and cross themselves. Prendergast waved them back and himself stepped forward with a trowel. He scraped away carefully as the sun rose higher and the exposed earth started to steam gently. Before long he had cleared around a bone that I verified as a human femur. Shortly thereafter, a whole intact skeleton lay exposed, and beside it a slab of slate marked with an intricate pattern. He picked this up, trembling with excitement, and gently cleaned its surface with a soft brush.

"Gentlemen, this is the very evidence I was seeking. This barrow dates from relatively recent times: the fourteenth century. But the map shows the location of much older graves. And they are just where I suspected: in the forbidden wood behind the counting-stones, which the locals for some reason call the Devil's Bowl. Father, I remind you of your promise: I have the evidence I needed to prove that a dig there will be fruitful!"

Back at the counting-stones, he and the men hacked at the dense wall of greenery with their shovels. They soon broke through into a clearing, but our excitement was short-lived, for ten yards further on was a dense thicket of bramble that looked to be impenetrable. Prendergast made a rueful face.

"I think we will have to send back to the house for some scythes and machetes. We will not get far trying to beat this down with spades."

At that moment there came a commotion in the long grass nearby. It looked as though some quite large animal was leaping toward us in pursuit of prey. Prendergast drew back in alarm as the invisible creature almost struck him, but a moment later, we saw that the "creature" was actually a pair of rabbits (or perhaps they were hares; my country knowledge is really abysmal) loping alongside and around one another, evidently in courtship. They leapt on toward a place where the thicket was apparently impenetrable and, to our surprise, disappeared through, revealing that at that point it was a mere curtain of grass.

Before I could react, Dodgson sprang forward and clambered up to the place where the rabbits had vanished. For a moment he was silhouetted against the sky. Then suddenly, with a cry, he disappeared from view. I scrambled more cautiously after him, and an extraordinary sight met my eyes.

I was perched on the edge of a grassy crater or bowl some two hundred yards across and thirty or forty feet deep. It was studded with rings of mushrooms, the kind that ignorant people call fairy circles, and large white stones lay everywhere at intervals of several yards. Dodgson had tumbled onto his back a few yards down the slope and was making no attempt to stand up. I clambered down to him, but he appeared unhurt.

"Follow a white rabbit, and tumble down into a wonderland!" he was muttering dazedly, although I was unsure whether he was talking to himself or to me. Prendergast and the others joined us, and we raised the Reverend to his feet and stood gazing about.

"This is it!" said Prendergast excitedly. "This is the ancient burial ground that was used in earlier times. Each of these stones marks the resting place of a king or queen of Arthur's ancient lineage. In a few minutes, we shall know which legend is true. You see, there are two differing stories about how these ancient royals lived and died. Both versions agree that a man must marry to become king, so as to rule with the benefit of both male and female insight, and that king and queen were in due course buried here."

Prendergast eyed us intently as he continued. "Now, one version has it that their marriage was for life, in the modern Christian tradition, and that each was buried when they died of old age or natural causes. But a rival story is darker. It maintains that the king would every seven years take a new young wife, divorcing the old. When the king eventually died, his current wife was immediately killed by beheading, to leave the way clear for the succession. To resolve the matter, we have but to dig up a grave containing a female body and verify that its spine is intact."

I bent down and peered at one or two of the white stones which were the grave-markers. But none seemed to have any writing or other indication upon it. "How can you tell which type of grave is which?" I asked. From behind me there came a harsh laugh: I turned and saw that the Mage had managed, despite his age, to follow us into the bowl.

"Quite so!" he said triumphantly. "I gave my son permission to dig up a grave here—one grave only—on condition that there would be a greater than even chance of finding out the truth. Now by either legend, the number of male and female skeletons buried here will be equal. If you dig up a grave at random, there is an even chance it will turn out to contain a male skeleton, which will tell you nothing. By my edict, you are not allowed to proceed."

Prendergast grimaced, but I could tell by the expression on his face that he saw no way to dispute his father's argument. At his suggestion, we spread out and started to comb the surface of the bowl, though what we hoped to discover was not clear to me. I found myself drawn to a corner where a huge oak tree stood. Its roots had disturbed the ground about it, one great shaft running deep between two of the marker stones. My eye caught something bright in the grass. I bent down and picked up a gold ring, its surface miraculously untarnished, made in the form of the Norse Midgard serpent that lies circling the world with its tail in its mouth. My shout brought the others running.

"It is a woman's ring, a queen's ring," Prendergast shouted. "Show me exactly where you found it."

Alas, I was forced to point to a spot exactly halfway between the two nearest stones.

"You cannot possibly be sure which grave it came from. You do not know which is the female one," said the Mage firmly. Dodgson was about to speak, but Prendergast raised his hand.

"Thank you, Reverend, but I can solve this one for myself. Father, we know that one of these graves definitely contains a woman's remains. The other has an even chance of being man or woman. So if we dig up one grave, the chances that it contains a female skeleton are three in four. By your edict, we may proceed."

The dig took some time, for we proceeded with both caution and reverence. We unearthed the legs, then the pelvis. I was able to identify the pelvis as definitely female, and Prendergast gave a cry of triumph. But as we excavated toward the head, we fell victim to an extraordinary piece of bad luck. A root of the tree had pushed just past the top of the rib cage, and the ground became very wet at that point. Beyond the root there was no further sign of bones. The neck bones and skull, intact or otherwise, were gone.

"I am afraid that the combined work of the tree roots and an underground stream has long since carried that part of the skeleton away, to be scattered and destroyed," I said when it was clear there was no further hope.

Prendergast flung his trowel on the ground in fury. "What an incredible mischance. Really, the gods themselves seem to be against me," he shouted blasphemously. "Father, in the circumstances, may I open the second grave?"

The Mage smiled maliciously. "Of course not," he said. "The ring could obviously have come from this female skeleton, so the chance that the remaining grave is a woman's is again only one in two. I cannot allow you to proceed."

We resumed our seemingly futile search of the bowl. I was devastated that my potentially useful discovery should have led nowhere, and Prendergast must have been feeling far worse. But suddenly there came a cry from the Reverend.

"I have it. Really, I am almost tempted to shout Eureka! The chance that the second grave by the tree root contains a woman is not one-half. It is two-thirds."

The Mage looked at him scornfully. "One-half to two-thirds," he said savagely. "That seems to be your theme song,

Reverend, but I am afraid I will not take your word for it. Surely we know nothing about the sex of the second grave."

Dodgson made no reply, but bending to the ground, he picked up a white pebble and a black one. Then he turned to me with a smile.

"Doctor, would you be so good as to lend me your top hat?" Although somewhat baffled, I gave it to him.

"Let me once again demonstrate the point as a children's game," he said. "I will shuffle these pebbles in my hand and place one at random into the hat. The other I discard without looking at it. So the hat contains a white or a black pebble, with equal probability." We nodded.

"Now I pick up a second white pebble"—he did so—"and place it in the hat. I toss the pebbles around so I cannot tell which is which.

"If a white pebble denotes a female skeleton and a black a male, I have created a puzzle equivalent to that of the graves by the tree. One is definitely white—that is female. The other is black or white—male or female—with equal probability. Now I take out a stone. I am in luck—it is white." He held up the pebble. "But unfortunately, the stone is mute as to how it met its end." He flung it down. "Now, given that the first stone was white, what is the chance that the second stone is also white?"

He paused. I felt there was something oddly elusive about the problem but was unable to put my finger on it. The others looked equally baffled. At one moment I convinced myself that the probability was only one-third, because we had already used up one female stone, so to speak. Or was it one-half after all?

Dodgson produced a sheet of paper. He pointed to the tree above us.

"The best way to illustrate the possibilities is by drawing a branching tree. I call it my many-worlds tree." He began to draw.

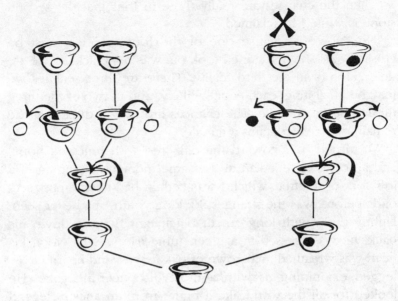

The Many-Worlds Tree

"At the start, we have one version of the world, with a hat that is empty. Now I place in it a stone that may be either white or black, and so we have two versions of the world—two potential worlds in which subsequent events will unfold differently. I add my second white stone, the same in each world, so there are still only two potential realities. Then I take out a stone at random. This may be either the original stone or the second one, so our two worlds fork into four realities."

He counted them across from the top. "In the first reality, I discard the original stone. The second was white, so the remaining stone is white. In the second reality, I discard the second stone, but the original was white. So the remaining stone is again white. In the third reality, I discard the black stone, so that remaining is white. In the fourth reality, I discard the white, and the remaining stone is black."

"Then the chances are really three in four that the second stone is white," I exclaimed.

"No, Doctor, because one of the four realities must be crossed out. The first stone I took out was *not* black, so we are definitely not in the third reality. There are three realities we may be inhabiting, each equally likely, and in two of the three the last stone is white. The chances are two in three that the remaining grave contains a female."

We turned back toward the oak tree with renewed hope, and all of us gave a start at the same moment. Squatting in a low fork of the tree, which I was certain had been empty seconds before, was the strangest-looking youth I had ever seen. Lightly built, with long, uncut black hair flowing down his back, he was dressed in a green tunic. He was barefoot, his head was wreathed in a crown of oak leaves, and he sat cross-legged examining us with a most disconcerting gaze. He looked for all the world like a figure from an ancient legend made flesh. Then Prendergast gave a snort of laughter.

"It is only Wu," he said. As we walked toward the lad, he continued in an undertone. "Wu is a half-wit boy whose mother died in childbirth. No one adopted him, and he grew up quite wild. The village people have felt guilty at their neglect, and many leave gifts of food or clothing for him on their doorsteps at night; these are always gone by the morning. He has become quite a local legend, but he is harmless enough."

"What are you doing in Wu's place?" asked the boy in a lilting voice. The Mage looked about to make a rough reply, but to my surprise Dodgson stepped forward.

"We are only visiting, we mean no harm," he said. "We want to dig in the earth a little, to know about things that happened in olden times. Then we shall go and leave you in peace. You have my promise."

Wu sat motionless for some seconds and then nodded. Under the youth's gaze, we dug up the second grave. It was indeed a woman, and I am glad to say that Prendergast's fears were proved baseless. I could not determine the cause of death, but the skull and neck were intact, and traces of arthritis indicated that she had probably been of a good age when she died. The dark local legends could be discounted.

The next few days were passed in a rural idyll the like of which I, nowadays a hardened Londoner, have never experienced. Days were spent walking in the enchanted forest, now devoid of dark pagan overtones for both myself and Prendergast. It seemed a fantasy land: as we walked, curtains of leaves in different shades of green sometimes gave us the illusion almost of being underwater. For all the signs of modern life we saw, we might have been in the greenwood of a thousand years before.

There were one or two discordant notes. Once I revisited the Devil's Bowl to see what looked like a freshly dug grave, and wondered whether Prendergast was continuing his excavations in secret, without either his father's knowledge or mine. And I found the child Wu tragic: neglected for so long that there seemed little hope that he would ever join normal human society, despite some patient efforts by Dodgson.

In the evenings, I tried to preserve a sense of reality by writing long letters to Holmes, describing all that I had seen in the strictly factual way that would meet his approval. Nevertheless, such was the magic of the place that I really came to feel I was not only in a different world but also in a different time from that which I had recently inhabited in London. Thus, when the lounge door opened on the fourth day to reveal Holmes and Lestrade, I gaped as though I were seeing a pair of Martians. After a few seconds, it occurred to me that a garbled account of our doings might have reached Lestrade's ears.

"Gentlemen, if you thought that skeletons must be a police matter, you are some twelve hundred years too late," I said at last. Lestrade shook his head.

"A graveyard is a graveyard, and never closes for good," he said sinisterly. "You are not the only people to have been digging hereabouts. We have good reason to believe that the remains of a moneylender called Robinson were interred at the place you have been investigating. Not in antiquity but more like last Wednesday, August second, Robinson having been seen alive on the Tuesday. That's August 1900, not A.D. 700 or wherever your thoughts may have been recently, Doctor. And I have here an exhumation certificate. It is signed by an Assizes judge and takes precedence over any objections by this Mage of yours."

Holmes nodded somberly. "He is right, Watson. I am certain we will find Robinson's remains there. I know it may upset your hosts, but there is no alternative."

I sighed. "Well, I suppose my holiday from the present century had to end sometime. I think I can show you where to dig. I have recently noticed fresh-turned earth."

We must have seemed a bizarre procession to any observer: a parody of a funeral cortege, with the Mage leading the way and thumping the ground at intervals with his great staff, accompanied by four of his retainers looking no less sinister for carrying shovels rather than halberds slung over their shoulders. Holmes, Lestrade, and Prendergast followed, with Dodgson in his clerical dress bringing up the rear. As we came to the burial ground, I trotted ahead to indicate the new plot. The men were ready to set to with their shovels, but Holmes held up his hand. For a few minutes, he ranged hither and thither about the site with eyes fixed on the ground. Presently he gave an exclamation. He had found a second freshly dug plot, which we clustered around in some bewilderment. Holmes continued to ignore us, and shortly he located a third. Then he paced about the rest of the site thoroughly without finding another.

"Are we looking for three corpses, then?" I asked.

"I think not. They have just been trying to buy time. The guilty man will be out of the country if we are not quick enough with a warrant," said Holmes.

"Well, this won't help them," said Lestrade cockily. "We'll have all three up in a jiffy." He gestured to the men, but the Mage held up an imperious hand.

"You have showed me one exhumation certificate, not three," he said. "And one excavation I shall permit, but no more."

Lestrade turned puce. "There is only one body we want, *sir*," he said, making an evident effort over the last word. "The other two will turn out to be empty holes. Time is vital."

That was an imprudent tone to take with the Mage. "Show me three certificates, and you may dig three holes," he said ponderously.

We all hesitated. One chance in three was not good odds, but we could hardly fight the Mage's men physically. It was obvious from the expressions of both Lestrade and Holmes that they did not think there was time to go for either more warrants or police reinforcements without jeopardizing the chase. But suddenly there came a peal of laughter from the edge of the trees.

"Wu saw! Wu saw the men bury the dead one."

We all turned toward the half-feral boy. Dodgson stepped toward him, holding his hand out in a gentle gesture.

"Where, Wu? Which new hole?" he asked.

But the boy shook his head fearfully. "Wu promise not tell. They say they do bad things to me if I tell. I promise not tell," he said, his voice rising in pitch.

Dodgson hesitated a moment, then half-smiled.

"Wu, tell me then which new holes are empty. Which have nothing bad in?" he said.

Wu hesitated. I thought he was about to comply. Then suddenly he cowered back as though struck. "You trick Wu!" he shrieked. "If I tell you not hole, and not hole, it is same as telling you hole with body. I not do it."

Dodgson nodded immediately. "I take it back, Wu. Do not tell me. I do not want the bad men to hurt you.

"But, Wu, could you tell me *one* empty hole? Surely that is not breaking your promise."

Wu considered. "If I tell you one empty hole, you still not know body hole," he said slowly. "You nice man, you give Wu chocolate. Very well, I show you one empty hole."

I saw that Dodgson was doing his best for us. The odds would be shortened from one in three to one in two. But to my astonishment, as Wu moved forward, Dodgson held up a hand to stop him. He walked round peering at the three holes. I looked intently—as did Holmes!—but neither of us seemed able to see any significant difference. Nevertheless, Dodgson moved to stand decisively by the central hole.

"I do not want to know about this one," he told Wu firmly. "But tell me one of the other two holes that is empty."

Wu pointed to the uphill one of the two remaining and then bounded back into the forest. Dodgson pointed to the third, downhill hole.

"Dig there," he said.

I saw a look of intense concentration on Holmes's face. Suddenly he smiled and nodded. "It's a two-in-three chance, Lestrade. Let's take it."

Lestrade looked quite baffled, but gestured the Mage's men forward. They dug energetically. One of them winced as his blade struck something. Moments later, we were looking at what must be the last mortal remains of Robinson, late moneylender of London. Holmes bent over the body and extracted a wallet and some papers from the clothing.

"Let's get to the police station, Lestrade. I'll tell you the warrants to make out as we go," he said, and the two men left at a run.

I came down to the drawing room for sherry before dinner to find that only the Reverend was present.

"I have been thinking about it as hard as I can," I said apologetically. "But I still do not understand why Holmes thought the chance of finding Robinson's body in the lowermost grave was two-thirds. If you were no more able to distinguish a dummy grave than I was, surely the chance with only one ruled out would be one-half. On the other hand, if you had spotted one of the graves as fake and got the boy to rule out another, then you were certain of success. How can the chance possibly have been two-thirds?"

The Reverend smiled gently. "You would have understood if you had not slept on the train, for by chance I was playing a very similar game with those schoolgirls," he said. "Let me re-create it for you."

From somewhere on his person he took three empty match-boxes and a supply of chocolates. In one box he placed one of the chocolates. Then he shuffled the boxes behind his back and placed them on the table. "Try to pick the box with the chocolate," he said.

I pointed randomly at the leftmost. He immediately opened the central one to show that it was empty.

"Now, would you like to change your choice to the remaining box or stay with your first choice?"

"I will stay with my first choice."

He opened my chosen box; it was empty. Then he opened the rightmost and ate the chocolate himself.

"Do you think that was pure bad luck, Doctor?"

"Well, of course. The chocolate had a one-in-three chance of being in any box. The chocolate did not move about, so showing me that the middle box was empty surely gave me no information about which of the remaining ones it was in! Change my choice or not, surely the odds were the same." I sighed. "I suppose now you are going to draw one of those infernal many-worlds trees."

He shook his head. "No. This kind of game, where there are in a sense two players in competition, is best shown a differ-ent way. It will be obvious to you once I make you the quiz-master." He passed me the three matchboxes and several more chocolates from his pocket. I placed a chocolate in one of the matchboxes, then shuffled them carefully so that I knew the winning box, though he did not. He promptly pointed to an empty box as his first choice. I thus had no choice: of the other pair, I opened the empty one.

"Would you like to change your choice?"

He smiled. "Thank you, yes." He took the remaining box and ate the chocolate. "Please pose the problem again."

I had acted as quizmaster several more times, and had watched him win most of the chocolates, before I saw it.

"Ah, how obvious. If your policy is to stick with your original choice, you win only one time in three."

"That is indeed obvious."

"But two times out of three you will pick an empty box. In that case I have no choice. I am constrained to open the only empty one of the other two. Then changing your choice guarantees you a win two-thirds of the time. There is no mystery at all!"

He sighed. "In a sense there is not, Doctor, though to my merely human intuition, there is still something a little spooky about it. It is fairly obvious that if the quizmaster has to open one empty box at the very start of the game, he increases the player's chances from one in three to one in two. But it is indeed bizarre that allowing the player to place a *purely random* constraint on which box the quizmaster is allowed to show increases the value of the information he gets!"

The Reverend smiled mischievously. "You do not find that startling, Doctor? Ah well, your mathematical intuition must be different from mine. Perhaps it is superior. Now, let us proceed in to dinner before we ruin both our appetites."

The following day Holmes, Dodgson, and I shared a compartment as we traveled back east toward Oxford and London. Holmes and I chatted, but the Reverend scribbled frantically in his notebook. At length I asked him what he was writing.

He beamed at me. "You remember I was debating whether to write a children's fantasy or a book of mathematical puzzles? Well, the recent stimulus has been such that I am quite resolved to do both. Your colleague Mr. Holmes suggested a way to overcome one obstacle that has deterred me: I am a shy man who would hate his name to be too much in the public eye. Mr. Holmes kindly pointed out the usefulness of a pen name."

He held up the title page of his manuscript.

I read aloud: "*Alice's Adventures in Wonderland,* by 'Lewis Carroll.' I wish you every success with it, Reverend."

6

The Case of the
Martian Invasion

THE YOUNG MAN GESTURED FRANTICALLY at me, without rising from his position lying flat between two outcrops of turf.

"For humanity's sake, keep your head down when you crawl," he hissed. "If the Martians see us now, they will have us, and Earth's last hope will be gone!"

There was so much conviction in his tone that I genuinely felt the hair prickle on the back of my neck. I almost found myself believing that if I were so incautious as to raise my head, the Martian death ray so vividly described by Mr. H. G. Wells would indeed destroy us in a searing blast of heat, as pitilessly as we pour acid to exterminate a nest of ants.

I waited alertly, but the signal to resume our tortuous progress did not come, and I found my thoughts returning to the start of the remarkable sequence of events that had led to my present predicament.

I had entered our living room to find my roommate alternately chuckling and shaking his head over a letter he had just opened. "I am glad to see you, Doctor," he said, "for if ever there was a client who is in reality more in need of your assistance than mine, this must be the one. Take a look at this!"

He tossed the letter across to me, and I read the following short but startling text.

Mertford College, Oxford
August 31, 1900

Dear Mr. Holmes,

I have the honor to be a graduate scientific student at the above institution, specializing in astronomy, but with a strong interest in applied physics, and the science of flight in particular. I have reason to believe that in the coming months and years great advances in this field will be announced, and that I have significant contributions of my own to impart.

However, with the turning of the century, my joy at the future has become overshadowed by a terrible fear. I have seen increasingly definite signs that the human race may be under threat from Powers almost beyond our comprehension. Incautiously, at first I discussed these matters quite openly, and now I have new evidence that I, personally, have become their target.

I do not know to whom I can turn for help—certainly not to the authorities. But perhaps you can advise me on a burden of dreadful knowledge that should not fall upon the shoulders of one man alone.

I hope to call on you at 9 o'clock on the morning of September 1st, and beg that you will postpone any other appointments. They can hardly be of greater importance than mine, for the fate not just of myself, but of the whole human race, may hang upon my imminent actions.

Yours most sincerely,

Alexander Smith

"The poor man, Holmes. I have rarely seen clearer evidence of incipient paranoia. I suspect it has gone beyond the point where soothing words and a sedative will put matters right. You really need the services of a specialist, but of course, I stand ready to give any assistance I can."

"That is good of you, Watson. For there is no time to engage a specialist: 9 o'clock is striking now, and I suspect that is Smith's tread I hear upon the stair."

A moment later we were shaking hands with a tall, red-headed young man with a look of confidence and authority about him that belied his years. A slight rigidity in the cheeks, a fixedness of the eyes—these were the only signs that betrayed his anxiety to my medical observation. Holmes waved him to a chair and, to my surprise, initiated his inquiries upon what seemed to me a side issue.

"Mr. Smith—soon to be Dr. Smith, we expect?—I gather you have some views upon the incipient science of flight?"

"Certainly I do, Mr. Holmes. Although my degree is in pure science, I can see that the engineering of the new century will become ever less empirical, and ever more dependent on mathematics and physics. I have made something of a hobby of applying the laws of probability and statistics to engineering systems. For example, I am fascinated by developments in both lighter- and heavier-than-air machines. I confidently expect that in the coming decades, we will conquer the atmosphere as surely as our modern ocean liners have already conquered the seas. You have no doubt seen pictures in the newspapers of that first ship of the air, the dirigible the *Graf Zeppelin,* which is even now under construction in Germany?"

We both nodded.

"It is an impressive project. Although I think that in the longer term, heavier-than-air flight is likely to be more effective. The key point will be achieving safety and reliability."

Already, I could perceive that the mind before us might be less than sound.

"I am well aware that heavier-than-air flight is theoretically possible," I said. "But I am sure you have noticed that the great governments of the world are initiating airship-building programs. Heavier-than-air flying machines, by contrast, are invariably built by wealthy private inventors regarded by their neighbors as eccentric. Does that not tell you something about the relative merit of the two approaches?"

Smith regarded me scornfully. "Not necessarily," he replied. "One key point is that the laws of physics confine an airship to traveling slowly in the lower reaches of the atmosphere, at the mercy of the weather. Other types of flying machine could travel much higher and faster."

"The question is one of safety, as you say," I said soothingly. "If an airship's engine fails, it does not fall from the sky but hangs perfectly safe, like a balloon. Why, it could even be taken in tow, if need be! Whereas a heavier-than-air machine must maintain an impressive speed merely to remain aloft. Gasoline engines are prone to failure: I have noticed how often these new-fangled motor cars and motor bicycles are to be found stranded by the side of the road, their chauffeur-engineers tinkering with them."

Holmes nodded. "Mr. Smith, I agree with the doctor's reasoning. Heavier-than-air flight is theoretically possible, but engine failure will lead at best to a crash landing, and with the incidence of failure inevitably once per hundred hours or so, ambitious flights such as trips across the English Channel will simply be too dangerous to contemplate. Even if you could in time improve engine reliability ten- or a hundredfold, it would still be too risky for most people."

The young man curled his lip. "That is just the area where I am making my own contribution," he said. "I agree that safety is vital. I have discovered a technique whereby any mechanical device, even an ambitious one such as a flying machine, can be made for practical purposes *infinitely safe*."

Holmes and I were silent, which Smith took as an invitation to continue. "I call my technique redundancy. The object is to construct a design where any one part can fail, or even any several parts, without affecting operation. For example, in the case of aircraft propulsion, we simply make every machine use not one large engine but several smaller ones, spaced apart along the wings. A good number of engines might be four. The aircraft will need less power merely to remain aloft than it does for its initial acceleration and climb to altitude, and we will give it some reserve margin in any case, so we may assume that only three, or perhaps even two, engines could provide enough power to take it on to a safe landing."

"Well, I suppose multiple failures could still happen," I said.

"They could, but it is most unlikely. If the chance of one engine failing in a given flight is 1 in 100, the chance of each of a pair failing is 1 in 10,000, and of all of a set of three failing just 1 in a million. You would have to be a pretty fair coward to decline to board such a vehicle!"

I flushed angrily, but fortunately Holmes interrupted before I could reply.

"I have reservations, but I can see your system has some merit—"

"Some merit!" cried Smith. "It is fundamental, absolutely fundamental. Have you not heard of the man who invented a perfect mechanical typesetting machine—except that it had so many parts to break down that it was hardly ever in working order? Have you not heard of Mr. Babbage, who has designed a computing machine that would almost be capable of thought—except, again, that the number of cog-wheels and so forth required to function perfectly together is alas so great that a working version remains a daydream? With the principles I have worked out, you can make a machine of pretty much any size and complexity, in the knowledge that it will never go wrong. It truly pushes back the frontiers of the possible."

"I suppose the sky is the limit," I said as enthusiastically as I could, recalling the importance of humoring the mentally unwell. But he glared at me.

"The sky is not the limit," he said fiercely. "Have you not heard of the Russian schoolteacher Tsiolkovsky and his proposed space-train? There are three known methods of flight. The first, the balloon or airship, depends on the static buoyancy of the air. The second, the airplane, depends on thrusting air downward to provide dynamic lift. But the third is the rocket, and a rocket requires no air. A single rocket cannot fly to the Moon, for chemical fuel has not sufficient energy. But if you took a stack of rockets, and fired each in turn, jettisoning the exhausted ones and adding cumulatively to the speed, there is no limit to how high you could go. You would need many stages, and the whole machine would be quite complex. Yet with my methods it could still fly successfully."

"To the Moon—really!" I cried, wondering whether he was genuinely mad or was merely pulling our legs. But he nodded quite solemnly.

"Yes—and that is undoubtedly what has brought Their attention upon me," he said. "I have long been a fan of fantastic fiction."

I caught Holmes's eye at this point, and each of us had difficulty keeping a straight face: clearly this was the first thing Smith had said that we could both unreservedly accept.

"I have read of Mr. Jules Verne's space-gun, and Mr. Wells's antigravity material Cavorite," Smith continued, "and been amused by these fantasies even while recognizing the inadequacy of their scientific content. Yet Mr. Wells's hypothesis of Martians cannot so readily be dismissed. For competent observers have seen lines on the Martian surface so straight and long that no natural origin seems plausible. It is guessed that they must be canals, implying the presence of intelligent Martians."

"No one has actually succeeded in photographing these canals," said Holmes sharply.

"No, but experienced astronomers have repeatedly described them. The turbulence of the atmosphere blurs the sharpness of a photographic image, which must be exposed for several seconds, whereas the human eye can snatch detail in the occasional instants when the air is still. But let us leave the canals for the moment. For I have much more irrefutable proof of the existence of the Martians. Oh, how I wish that I did not!"

For the first time, the young man's self-control seemed about to desert him. With an effort, he regained his composure. "The problem is that Mars is even at conjunction more than thirty million miles from Earth, and that obviously limits what we can see. Our own Moon, on the other hand, is but a quarter of a million miles away at all times.

"Now, it has long been known that, Mr. Wells's charming Selenites notwithstanding, the Moon cannot really be inhabited, for there is no air there to breathe. Nevertheless, it is plausible that we terrestrials may find some way to go there and explore its surface, wearing something like diving suits to walk about in. It occurred to me that if there really are Martians, and they are more scientifically advanced than we, they might very well have made similar expeditions. And because the Moon has no wind or rain, the traces of any such visit would remain forever, even if it took place thousands of years ago."

Our visitor's voice grew confidential. "So I borrowed time on one of the University's smaller telescopes and embarked on a paradoxical search whose real objective I obviously did not reveal, for I would undoubtedly have been thought insane. I was looking for evidence of Martians, not upon Mars but upon our own Moon. I photographed every square mile of her surface, and at different points of the lunar day, paying particular attention to the lines of dawn and dusk, when the long shadows cast can reveal details of the smallest objects. After many nights, I had spotted nothing unusual, when suddenly I came across—this!"

He had opened a battered briefcase as he talked, and with a dramatic gesture, he drew forth a large photograph and slapped it down upon our coffee table. Holmes and I peered intently at it. To me, it looked like a mere jumble of rocky terrain.

"Here is the eye," he said, pointing. "Here, the line of the jaw. The mouth is distorted but starts at this point—"

"Good lord!" I gasped, for suddenly a hideous and distorted face had sprung out of the landscape at me, just like one of those trompe l'oeil paintings.

"It is striking, is it not? It inspired me to continue my search. And soon I came across a second work of art, if such it can be called. Think of an octopus."

He put down a second photograph and pointed to it. This time recognition was immediate. An octopus-like outline was certainly visible, but with a bloated body topped by something like a face with gigantic saucer eyes. It looked indeed very like the drawing of a Martian that had recently adorned a poster advertising Mr. H. G. Wells's book *The War of the Worlds*.

"The Martian Face," said Smith solemnly. "The two features, the Martian and human faces, are not so far apart from one another on the lunar surface. The relative positions could suggest that the Martian is gloating over the suffering of its human serf. Doubtless images intended to strike fear into the heart of our race—but no one even recognized them for what they were, until now.

"Gentlemen, I was dreadfully shaken. I rushed to show the photographs to colleagues. But I was ridiculed and advised to take a rest from astronomy and concentrate on my proper studies. In a most sinister development, I was barred from further use of the university observatory—for my own good, I was told.

"I turned for solace to the pages of the Bible. I was surprised at myself, for I was never a religious man. But now I realize that a Power greater than ourselves, and greater even than the Martians, was guiding my actions. Gentlemen, have you a Bible in the authorized translation?"

I fetched the book, but Smith would not accept it from me.

"I would like you to see the message for yourself," he said, and directed me to a particular chapter, book, and verse. He pointed to a particular occurrence of the letter A. "Write down that letter, then every seventh letter thereafter, ignoring punctuation and the spaces between the words."

I did so. "It spells out 'Ares,'" I said.

"The Greek word for Mars. Continue onward."

"'Ares Comes!'" I exclaimed.

"Quite so, gentlemen. A phrase beyond mere coincidence, I hope you will agree. The most horrible significance lies in the position of the words. Chapter 19, verse 11, which we write as 19:11. Mars will attack in 1911. What could be plainer?"

"At least we have a decade to prepare," I said.

He shook his head. "It has since occurred to me to go through Chapter 19 of every book in the Bible, looking for similar patterns. I have uncovered many dozens of sinister phrases containing the word 'Ares,' but in verses of differing number, therefore pertaining to many different years in the coming century. Evidently we are to expect not a single battle, but a hundred years of war."

I knew that to cure the man, preserving the appearance of calm was vital, but I was by now quite startled. My next words were intended to reassure myself as much as Smith.

"But how could the Bible writers—or for that matter Mr. H. G. Wells—possibly know of an invasion that lay in the future?" I pointed out.

Smith shrugged. "Prophetic dreams, or nightmares, rather?" he suggested. "Or perhaps the authors themselves were quite unconscious of the influence that guided their hands. Although I cannot believe it is mere coincidence that the planet Mars represented the god of war in both the Greek and Roman cultures. But I am afraid we have no time for further scholarly study. I have spent the last year preparing a book detailing these revelations. However, I have also kept a careful eye on the newspapers, for if battle is to commence so soon, the first signs of action may appear any day. What I certainly did not expect, however, was that despite the vastness of the Earth's surface, a small village not a mile from me would be the first target. Last week I opened the local newspaper to find this."

He produced from his briefcase a battered copy of the *Oxfordshire Agricultural News.* The front page was occupied by a most extraordinary photograph. It showed a perfectly normal cornfield, except that near the center, a great circle of corn some ten yards in diameter had been flattened down in a clockwise pattern. The rest of the field was untouched. The headline read, "Mystery Crop Failure Baffles University Scientists."

"Perhaps an explosion?" I suggested. "Some pranksters, or someone with a grudge against the farmer, igniting a little gunpowder to cause a mystery?"

Holmes shook his head. "An explosion would leave the stalks pressed outward from the center, not flattened sideways. In any case, the boundary between the flattened stalks within the circle and the untouched ones outside would hardly be so sharply defined."

Smith nodded. "Pranksters are ruled out in any case. A man approaching the flattened area would have left a trail in the corn plain to see for days. No human agent can have entered the field the night it happened."

"Most singular. And your theory?"

"Some great machine, cylindrical in shape, came down gently on the field during the night. It must have been twisting slowly on its axis to flatten the corn in that pattern. It stayed for an unknown length of time and then left as silently as it had arrived. They are not willing to show themselves yet, but the first Martian scouting missions are already upon us."

It did indeed look just as if some massive object had come down on the field corkscrew-fashion.

"It can hardly be coincidence that this first landing took place within a mile of the first man on the planet to anticipate the Martian invasion. I can only conclude that they were somehow able to detect my telescope. It was pointed at the Face for long periods."

Even Holmes appeared at a loss for words as Smith continued: "Time is becoming very short. Even in the last few days, fresh corn circles have been appearing in the fields about Oxford. Cruder than the first, they suggest to me that the Martians are scarcely bothered about concealment any more, landing and taking off brazenly rather than in stealth. And now strange lights and noises have been reported in the night, just outside a nearby village. The Martians are coming, gentlemen! And I have decided it is my duty to risk all for irrefutable proof. Tonight I shall go armed to the site of these strange lights. Either I shall come back with an alien artifact, or maybe even the body of a Martian, as my proof, or else I shall not return at all."

He looked at my friend. "That is why I come to you, Mr. Holmes. You are known as a clever man. When you have worked out the odds against my evidence—the Martian Face, the Bible Codes—being due to mere chance, you will be convinced. You must then warn the world."

He rose from his chair, but Holmes held up his hand. "It is a brave mission. I believe there is less to fear than you think, but you could do with a companion."

Smith shook his head. "You must not put yourself at risk, sir: you must remain safe here in case I fail," he said resolutely.

I rose from my chair. "I will accompany you," I said. "Even though I have not Holmes's powers of detection, I am a reliable witness and a good shot with my revolver. Give me a place to meet, and a time, and I will be with you for tonight's reconnaissance."

I still did not quite believe in a Martian invasion, but in truth I had become almost irresistibly curious. The young man clasped my hand.

"Ten o'clock tonight, then, at the Miller's Arms in Sutton," he said.

"At first I was not sure whether he was a genius or a madman," I said to Holmes when Smith was safely out of earshot. "But now I am decided. Martians, indeed! Even his approach to engineering reeks of paranoia, designing as though he mistrusts every component he is working with."

My friend smiled. "The categories of genius and madman are not necessarily exclusive," he said. "There is a good deal of truth in the old cliché that you cannot have a great imagination without also occasionally being vulnerable to nonsensical ideas. Consider the example of Wallace, who deserved equal credit with Charles Darwin for formulating the theory of evolution. He came to believe in spiritualism, being taken in by simple tricks of table tapping and forged photographs, and is now ridiculed and ignored by the establishment. And most unfairly: just as the Christian tries to revile the sin yet love the sinner, we should not devalue a man's greatest work just because he has later fallen into error.

"Our young man has perhaps the makings of a great engineer. His concept of redundancy is most striking. The last century has been one of colossal feats of engineering, yet also of great tragedies: our boldness has had its price. The number who die annually upon the railroads is excessive, but it is dwarfed by the death toll of those who labor to build them and the structures and bridges they require. Those who work with machines do so at great risk to life and limb. Perhaps the new century needs to be one of more delicate, more subtle, yet safer engineering.

"It is worth a good try at curing his obsession, Watson. And I would take your revolver with you tonight. I do not believe in Martians, but as to the true cause of the events in Oxfordshire, I am as baffled as yourself."

The Miller's Arms turned out to be a rough country pub. Smith was not yet there, so I bought myself a pint of watery ale and engaged the landlord in conversation about the mysterious crop circles, seeking discreetly to draw out details. To my discomfort, several men sitting at the bar listened intently to our talk, and I was relieved when Smith arrived. We took seats at a table covered with puddles of beer, and he sketched upon it with a damp finger.

"Here is the pub we are sitting in. I place this ashtray to mark the spot where the light has been seen. There is quite a maze of small quarry pits between here and there, but I will be able to guide us. The original crop circle was in the field to our south, here"—he used the rim of an empty beer glass to make a damp circle on the table—"and subsequent ones have appeared here, and here. Do you have your revolver with you?"

"I do. Let us be on our way."

The pub was in an isolated spot, and soon we were making our way through pitch blackness, for there was no moon. Presently Smith gestured to me. "Here come the quarry pits. Follow my path, and be careful to take no misstep."

I followed him over hillocks and along ridgeways between deep but grass-covered pits that must have looked charming in the daytime. The place was like a miniature version of a Swiss landscape, with hillocks and valleys just a few dozen feet high and deep. It occurred to me that this was extraordinarily like the ground Mr. H. G. Wells described his first Martian cylinder as falling into. I tried to recall his thrilling description: the projectile had embedded itself in the wall of a pit. The lid was screwed off very gradually. Eventually a tall antenna was seen rising above the site; it turned out to be the projector for the dreaded Heat-Ray. And then came a shape as tall as a church steeple, but moving—the first of the terrible Fighting Machines!

It was at that very moment that a strange creaking sound came to our ears. We both stood stock still. It came again, and again, with great regularity. There was no doubt in my mind: this was the sound of some massive metal contrivance turning very slowly!

Smith got down on his hands and knees, gesturing to me to do the same, and we crawled forward between the outcrops of turf for whose concealing presence I was now most grateful. Presently I peered forward, and my heart almost stopped. Before us was a great dark shape silhouetted against the stars. I estimated its height to be at least forty feet. It could have been a derelict building, but it was unmistakably moving, the top edge eclipsing the stars, sweeping not only sideways but downward. The eerie sound came again. I tried to discern the object's motion, but it seemed to be bending and straightening in one position, rather than progressing along the ground.

We were both frozen in place when we had an even greater fright. To the left of the building, a tall, thin pole at least twenty feet high, with some object at the top, was wobbling about. And the pole was unmistakably moving toward us, though in a jerky and erratic fashion. I could hear my companion trying to suppress the chattering of his teeth.

A light shone out from the top of the dark object. For a moment I thought of the Heat-Ray, but no, this was merely illumination. A moment later came the last sound I expected to hear.

"Who's that in the dark?" said the voice of a young woman sharply, and then louder: "Father! Shine the light down, there is somebody here."

The light swung toward us, revealing a tall girl with straight blond hair. In her hand she held a long bamboo pole with a gauze bag at the end: a collector's butterfly net. Behind her, the glow showed an old derelict windmill. Its sails turned slowly in the light breeze, and there came from it the regular creak of unoiled bearings.

We stood up, dusting ourselves down with as much dignity as we could muster.

"May I ask what you are doing, creeping up on me like that?" said the girl coldly.

"We came to investigate lights and sound by the quarries," I said sheepishly.

"It is public land, and we have every right to be here. But if you must know, we are collecting moths. My father is a keen lepidopterist, and shining a light from the top window of the old mill brings them for miles. He has a theory that the color of the moths is changing in response to the industrialization of the landscape, demonstrating evolution in action."

We made our way back without any attempt at concealment.

"Try to see the funny side of it," I said to Smith. "We are all destined to feel foolish now and then, but at least we can be relieved that they were not Martians after all! Come, if we hurry, we will be in time for a nightcap at that pub."

We reached the door just as the bell for last orders rang out inside. But we were stopped in the hallway by an embarrassed-looking man with sun-browned face and callused hands.

"See 'ere, I hope there won't be no trouble about my corn."

"This is Farmer Wilson, on whose field the first crop circle appeared," explained Smith.

"You see, it's like this," the man continued awkwardly, shuffling one foot. "You know them new-fangled water-sprinklers, the ones that go round and round as they spray?"

We nodded. I had of course seen domestic versions on many a London lawn.

"Well, I was using one of those, but a bit unorthodox-like. We've 'ad terrible trouble with weeds round 'ere, real terrible. And I got to thinking, if I use one of these chemicals to treat it, it'll do no 'arm really, and who's to know?"

"Using weed-killer in this area is not allowed because of the runoff into the Oxford drinking reservoir," Smith explained.

"Well, it was only a little bit," said Wilson, not meeting my eye. "I fastened a canister of the crystals onto the farmyard tap. Then I ran the 'ose out toward the cornfield. I didn't actually step into the field, so as not to trample the corn. I just threw the sprinkler out as far as I could, the 'ose trailing behind it.

"At nightfall, when no one would see, I turned on the tap for an hour or so. At dawn, I went out to the edge of the field and reeled in the hose. But the weed-killer was too powerful, and it 'ad brought the corn down round it along with the weeds."

"Of course!" I said. "The clockwise rotation of the sprinkler put more water and poison on one side of the stalks than on the other, so they came down in the pattern we saw."

"I only done it the once, and I won't never again," said Wilson.

"What about the other circles, then?" I said sternly.

"I 'ad nothin' to do with them, honest. They're not even on my land. Perhaps some of my neighbors got the same idea."

"Well, if you promise not to spray any more poison, perhaps we can overlook the matter," I said sternly, and we passed on into the bar, just in time to purchase pints of beer before time was called. As we quaffed, a roughly dressed younger man came up to us. I recognized him as one of the bar loafers of our earlier visit.

"We didn't mean to cause no trouble," he said in a whining voice. "It was just a lark on the way back from the pub, see?"

"I think you had better tell us about it."

"Well, nothing much ever happens round here. Then Farmer Wilson's circle got in the newspapers, and men was coming and paying to take photographs, and all them scientists, and all. So we took a stake and a bit of rope and a double stepladder. Then we climbed into a couple of the fields by the road, using the stepladder so we didn't trample the corn at the edge. We bashed the stake in and tied the rope to it and walked round in a circle; that's all there was to it. Come daylight it didn't look as neat as Wilson's circle, though. Honest, it was only for a bit of a laugh."

I am nowadays a little old for crawling around on wet grass in the middle of the night, and the following afternoon at Baker

Street, it was while inhaling breaths of steam from a basin of hot water and lemon, a towel over my head and my slippered feet stretched toward the fire, that I delivered my report. Sherlock Holmes listened in high good humor.

"Capital, Watson!" he said when I had finished. "Your description of the making of the crop circles is much more satisfying than one or two accounts by imaginative scientists I have read. One claims that miniature whirlwinds might cause such circles; another blames them on plasma vortices. I do not think either of these men can have much practical experience outdoors. It takes substantial force to smash down corn like that, and if either small whirlwinds or plasma vortices of such strength were common in our densely populated islands, it is hardly credible they would go otherwise unnoticed."

"I think the young man is on his way back to sanity, Holmes. So I can perhaps count it a successful treatment in my capacity as a doctor, although it is one of the more unusual house calls I have made. However I still could not explain his coded messages in the Bible, or his lunar pictures, so I asked him to call again here, in the hope that I could enlist your help to make the cure complete."

"I shall certainly do my best. I shall now set out to pick up one or two props that may assist me."

Smith arrived shortly thereafter, and Holmes seated him by the side table.

"The question of the crop circles and the night lights of Oxfordshire has now been cleared up," said Holmes, "but I have some further good news for you." He laid a stack of photographs on the table. "Here are some more shots of the Moon. In this one you can see a face peering from a crater."

"Why, I certainly can. Which part of the lunar surface is this?"

Holmes ignored him. "Here is a pyramid. And here the silhouette of a large animal like a wolf or a tiger."

"I see them."

I could not understand Holmes's motives at all. He seemed determined to make my patient worse, if anything! But suddenly he banged his fist down on the table. "These are not photographs of the Moon!" he thundered. "They are of some clay pits near Saffron Walden. An archaeologist friend of mine took them to record the location of shards of pottery found there: he kindly lent them to me for this purpose. The human eye is such a zealous detector of patterns that you can see such images almost anywhere if you try hard enough. That is why psychologists ask patients to inspect the inkblots known as the Rorschach test. As a child, were you never frightened by a strange person or animal in your bedroom, only to find that your own clothes draped over the bedside chair had suggested the sinister outline? Have you never started at a shadow, walking through a wood? If you take thousands of photographs of the surface of a body like the Moon and inspect them from different angles, there will be hundreds of suggestive shapes visible. Once you notice a particular one, and point it out to others, everyone agrees upon it, and it becomes the Lunar Face, or the Martian Face, or whatever, even if better examples are actually to be found elsewhere.

"Good grief, even given a sample of one picture–the whole Moon, the only astronomical body whose surface detail we can see with the naked eye–do we not see the Man in the Moon? He was probably pointed out to you in your cradle!"

Smith hesitated, then seemed to rally. "There is one part of my evidence that you cannot dismiss so easily," he said. "The code messages I found in the Bible. Consider that warning: Ares Comes!"

Holmes nodded. "I have considered it most carefully. I think that at the heart of your perceptual difficulties, Mr. Smith, is an inability to appreciate the vast number of permutations and combinations that can potentially be performed on a given set of data."

"On the contrary, Mr. Holmes, I have worked it out mathematically. That message is 9 letters long. There are 26 letters in the alphabet, so the chance of that message occurring randomly is 1 in 26 to the power 9, or less than 1 in 5 trillion. And I found not just one, but many similar messages. That cannot possibly be due to chance, in the mere few million letters that make up the Bible!"

Holmes shook his head. "First, you must remember that the letters of the alphabet do not occur with equal frequency. The E occurs most often, about once in every 8 letters, then T, once in 11, and so on. Because the letters that occur disproportionately often in the starting text are also more likely to be valid in a random word, the chance of getting real words from a formula such as 'every seventh letter' are much higher than you might think."

Holmes crossed to his library shelves, drew down a well-thumbed book on codes and ciphers, and opened it to a table of letter frequencies in the English language. "Consider the word 'Ares.' A occurs once in every 12 letters, R once in 17, E once in 8, and S once in 16. The chance of getting this word by starting at an arbitrary point and using an arbitrary formula is thus about 1 in 26,000.

Letter Frequencies in English

Letter	Percent	Letter	Percent
a	8.2	n	6.7
b	1.3	o	7.5
c	2.8	p	1.9
d	4.3	q	0.1
e	12.7	r	6.0
f	2.2	s	6.3
g	2.0	t	9.1
h	6.1	u	2.8
i	7.0	v	1.0
j	0.2	w	2.4
k	0.8	x	0.2
l	4.0	y	2.0
m	2.4	z	0.1

Holmes's eyes narrowed. "That may not sound high, but if you adopt the right system, you can find occurrences easily. You probably looked for each A, and then on until you came to an R. If the R was, say, eight letters beyond the A, you looked eight further letters for the E, and then if successful for the S. Your chance of generating *Ares* from a given starting A was nearly 1 percent. You would find several occurrences on every page! If you continued the sequence and it happened to generate any consecutive word, you had an apparently significant phrase. And if you found no valid following word, you no doubt tried for a preceding one—am I right? Hence it is totally unsurprising that in Chapter 19 of almost every book in the Bible, you found some apparent reference to Mars."

Holmes sighed. "Someday," he said reflectively, "someone will no doubt build a more powerful version of Babbage's difference engine and type the whole Bible onto punched cards, followed by a dictionary and a list of the names of every prominent person historical and contemporary, and then set the machine to search for messages. Trying every possible starting letter, and every reasonably simple formula for generating a string of letters, it will make billions of attempts and will inevitably find thousands of phrases containing every kind of prediction. And anyone who selects the more meaningful-sounding parts of this gibberish, and prints a book of it, will be able to 'prove' that the Bible—or indeed any other large text—is an amazing work of prophecy."

Smith stood up. "You have made your point, Mr. Holmes," he said. "I am cured. Now, how much do I owe you?"

"Nothing, Mr. Smith. Watson and I have been compiling a little list of common fallacies and errors involving logic, number, and probability, and your contribution has been most fruitful."

He ticked off points on his fingers. "First, you showed us how the human eye and brain can detect pattern where there is none. It is understandable design by evolution, for it is better to be frightened by ten shadows than to overlook one actual tiger, but it often trips us up in modern life.

"Second, there is the fallacy of retrodiction—conducting a blanket search of a great number of possibilities, coming up with some apparent pattern or message, and claiming subsequently how unlikely it is to get just that message in just that position. It is more often done by numerology: measure every possible dimension of the Great Pyramid, say, in every system of units known to you, and then try dozens of possible numerical combinations of the results to see whether any of the numbers that emerge seem significant, such as being a famous year in the Christian calendar. But your Bible messages have that beat all hollow. That leads us to a related problem. Before you go, Mr. Smith, may I counsel you on your novel notion of redundancy in engineering?"

The young man sat down resignedly. "I suppose you are going to tell me that is all nonsense too!" he cried ruefully.

"No, I think you are on to something promising. But you must be careful how you calculate the safety of your designs. Let us consider your example of a multiengined aircraft. Suppose that for a given flight, there is a 1 percent chance of any individual engine failing and that the airplane can fly with three engines, but no fewer. How likely is it to crash?"

"Well, the chance of each of a pair of engines failing is only 1 in 10,000."

"Yes, but there are four engines available to fail. That worsens the odds rather."

Smith scratched his head. He produced a used envelope from his pocket and on the back made a sketch.

"I see what you mean. There are four ways for one engine to fail, so a harmless single failure will occur on 4 flights in every 100. But there are six ways in which a pair of engines can fail, so the chance of that is 6 in 10,000. There are also four ways for three engines to fail, 4 chances in a million, and one way for all four engines to fail, with a chance of 1 in

$$\overline{XO\overset{\wedge}{\square}OO} \qquad \overline{OX\overset{\wedge}{\square}OO} \qquad \overline{OO\overset{\wedge}{\square}XO} \qquad \overline{OO\overset{\wedge}{\square}OX}$$

$$\overline{XX\overset{\wedge}{\square}OO} \qquad \overline{XO\overset{\wedge}{\square}XO} \qquad \overline{XO\overset{\wedge}{\square}OX} \qquad \overline{OX\overset{\wedge}{\square}XO} \qquad \overline{OX\overset{\wedge}{\square}OX} \qquad \overline{OO\overset{\wedge}{\square}XX}$$

$$\overline{OX\overset{\wedge}{\square}XX} \qquad \overline{XO\overset{\wedge}{\square}XX} \qquad \overline{XX\overset{\wedge}{\square}OX} \qquad \overline{XX\overset{\wedge}{\square}XO}$$

$$\overline{XX\overset{\wedge}{\square}XX}$$

Some Unlucky Aviators

100 million, but those possibilities are almost negligible by comparison."

"Your aircraft will crash once in 1,650 flights, near enough. You might like to consider using more engines. Suppose you use six engines, and the aircraft can at a pinch fly on four."

Smith started drawing more diagrams, but Sherlock Holmes coughed. "There is a rather more systematic way of working through the possibilities, or your paper consumption will become rather large. Imagine you are the flight engineer, keeping a log of which engines fail, and in which order. The first engine to fail can be any of the six. The second will be one of the remaining five. The last must be one of the remaining four."

"I see what you mean. So there are $6 \times 5 \times 4$ ways—that is, 120 ways—in which his log, recovered from the wreckage, may read."

"That is correct, but you must distinguish between the possible number of permutations and the possible number of combinations. Permutations take account of order, whereas combinations do not."

"It reminds me of an old music-hall joke," I said. "An incompetent pianist denies he is getting the tune wrong. The audience heckles persistently, and eventually the pianist admits that, although he is playing the right notes, they may possibly be in wrong order!"

Both men ignored me.

"For example, if the engines numbered 1 to 3 fail, they may do so in any order," Holmes continued. "There are three choices for the first to fail, two for the second, and the third follows inevitably. Six sequences of events result in the same set of engines dead, in the end."

"So the number of combinations—sets of three engines—is $6 \times 5 \times 4$ over $3 \times 2 \times 1$," said Smith excitedly. "Which is twenty pairs: 20 chances in 10,000. The chance of such a thing is 1 in 500."

"Still rather high," I said.

Smith nodded. "But I feel sure I could design the thing to fly safely on three engines, so four would need to fail," he declared. "And the chance of that would be 1 in 1 million—the chance of a particular set of three failing—multiplied by the quantity of $6 \times 5 \times 4 \times 3$ over the quantity $4 \times 3 \times 2 \times 1$." He scribbled hastily on his envelope. "It works out to 15 in 1 million. Why, that is remarkably safe. Mr. Holmes, I am indebted to you for your help, and I feel more confident than ever of my multiple-redundancy concept."

Holmes raised a cautionary hand in a manner I knew all too well. "It is in essence an excellent idea. But I have a further warning for you, young man. When making claims as to the safety of your products, be very careful to distinguish independent probabilities from those that are not independent. For example, on this logic you could claim that if you built a four-engined airplane able at a pinch to fly on just one engine, each of which is 1 percent likely to fail, the chance of total failure is only 1 in 100 million. It will never be that low."

"Why not?"

"Because the engines are never truly independent of one another. What are the most probable causes of failure?"

"One is bird-strike."

"Exactly. Birds often travel in flocks, do they not? The probability that an engine will suffer a bird-strike, given that another engine has just done so, is much higher than the probability of such an event on a randomly chosen flight."

"I picked a bad example, then. Mechanical failure will not follow such a pattern."

"Oh, yes it will, to an extent. Suppose all four engines come from the same defective factory batch? Suppose an apprentice engineer, servicing your airplane, makes the same mistake with all four engines? The probability that one engine will fail, given that another has just failed, is inevitably higher than in other circumstances."

"Confound it! I will make sure that each airplane has engines from different factory batches, and that different teams of engineers service each of the four engines."

Holmes smiled. "You are thinking along the right lines. I am not making fun of your system; it is potentially a very powerful one. But my last advice to you is to avoid the fallacy of thinking that real-life events can ever be quite independent of one another in probability."

Smith nodded stiffly and rose. "I suppose you have taught me a little humility, Mr. Holmes. In fact, I have lowered my sights from the sky for the time being. I intend to design something more practical: an unsinkable ship."

Holmes pursed his lips. "I believe there have been many such, most of which now lie at the bottom of the sea," he said. "How will you achieve this feat?"

"My design involves many independent watertight compartments, with sea-proof doors between them. The ship will continue to float if one, or even several, are breached. You are not about to suggest that the rivets will fall off widely separated hull plates at the same moment by chance, are you?"

"Not by chance. But ships commonly sink from running aground. If a reef scrapes a gash right along the side of the ship, your ingenuity will not save her."

Smith smiled. "My ship is intended for the London–New York run. You will not find many reefs in the middle of the Atlantic! You cannot dampen my enthusiasm for this project, Mr. Holmes. I shall talk to various shipyards later this week. I have already thought of a suitable name for the vessel, a name that will appeal mightily to shipping magnates."

"And what is it?"

"I shall call her *Titanic,*" Smith said proudly, and bowed himself out.

7

Three Cases of
Unfair Preferment

"You appear to be deriving considerable amusement from that magazine, Watson!"

I looked up. I could hardly deny the charge, for I had been chuckling for several minutes. "I am sorry if I disturbed you, Holmes. But it is not often one comes across a really ingenious new parlor game, and this one is excellent. The author invites you to imagine that three famous historical persons are in a balloon, drifting low over the sea. The wind is blowing them toward shore, but only slowly. They will all perish unless one is thrown over the side to lighten the balloon. For the game, you and two friends pretend to be the famous figures—the author gives the example of Newton, Julius Caesar, and Socrates. Each of you tries to be as persuasive as possible that he should not be sacrificed."

"And do you really believe people can be ranked in such a way, Watson?"

"Well, it is the principle on which Society operates," I said. "In principle, you could rank everyone in England. Starting with the Royal Family and the line of succession to the

Throne. Then the aristocracy, followed by leaders such as politicians and prelates. Then come the professions: doctors, lawyers, members of the clergy, and so on. Next the honest artisans, and finally the idle and the criminal classes."

"There would still be many people on each rung of your imaginary ladder, surely?"

"There would always be a way to decide," I said firmly. "For example, there are many marquises, but I would rank those who devote their time to administering their estates responsibly ahead of a black sheep like Whitebridge the gambler."

Holmes smiled. "It must be refreshing to see the world with your simplicity, Watson," he said. "Now, I could suggest a simple criterion that could be used to solve your balloon problem in every case."

"What is that?"

"Why, I would weigh them all, and overboard with the heaviest!"

"Really, Holmes, your prosaic approach could spoil the spirit of any game," I cried indignantly.

My friend's expression became sober. "I am afraid I have been reading of rather more serious matters, Watson." He pointed down at a pile of newspapers, all with headlines dominated by the extraordinary events in Sussex. I was recalled to sobriety.

"Yes indeed. It is almost unbelievable, Holmes, that such a man should have met his end in such a way," I said, shaking my head.

"I would have thought you were well hardened to murder by now, Watson, considering the length of our acquaintance!"

"To the bare fact of murder, certainly. But the victims tend to be people who have themselves led rather unsavory lives, or else people who are vulnerable, physically or mentally. Sir James hardly fits either of these categories." To illustrate, I held up one of the obituary pieces, and read the last few sentences aloud:

"In his early fifties, Sir James Vernon continued to dominate any opera in which he appeared by the sheer power of his presence. Perhaps the greatest tenor of our times after Caruso, even if he had never sung a note he would still be remembered for his abilities as a composer and violinist. Off-stage, his personal efforts as well as most of his wealth have been devoted to charitable pursuits, in particular the education of deserving young musicians from humble backgrounds.

"There appears to us no conceivable motive for the killing of a man who sought to help all around him. We do not envy the task of Chief Inspector Lestrade of Scotland Yard, who we understand has been assigned responsibility for finding the person who brought down an untimely curtain on this noblest of lives."

Holmes snorted. "I am surprised at your naiveté, Watson. I quite agree that most murder victims turn out to have had some flaw that contributed to their fate, but surely you are aware that a noble public persona may mask a quite different character. No doubt we shall soon be hearing of another side to Sir James. For unlike the writer of that article, I do not anticipate that even Lestrade will find his task too challenging. Sir James's photograph has often appeared in the papers, and he must have been one of the most easily recognizable people in the country. Wherever he went, whatever his doings, there will be plenty of witnesses who remember. I confidently expect that the case will solve itself."

As he was speaking, there was a knock at the door, which I opened to admit a telegraph boy. Holmes picked up a paper-knife and slit the envelope offered him. As he read it, his face assumed a somewhat comical expression.

"Well, you often neglect to inform your readers of the many times I am wrong, Watson, and this is certainly one of them. This is from Lestrade, and he appeals for me to join him in Sussex. Are you game for a weekend trip to the country? Capital! No need to rush your packing; I shall check the morning's mail first lest there is any urgent news on other fronts."

I returned to the room a few minutes later to find him frowning over a letter that I could deduce came from a woman, from the handwriting and the lavender paper on which it was written.

"A development in the Slingsby case, Holmes?"

"Hardly! No, this is from a very different correspondent. Do you remember Miss Catherine Lawrence?"

"Certainly I do, Holmes. Fiancée to the Marquis of White-bridge."

"Correct, Watson, but I fear not for much longer." He tossed the letter to me:

> Dear Mr. Holmes,
>
> Some months ago, you very kindly persuaded my husband-to-be of the futility of casino gambling. You explained to him that because there is always a percentage taken for the house, no system can in the long term win at games such as roulette. The following day, he pledged to me that he would never again enter a casino. He has kept his word, and has found respectable employment as private secretary to a former colleague of his father's. We are to be married in July.
>
> We were blissfully happy until last week. It was then that he told me of a new club in which he has been offered membership. It is a sort of gentleman's gambling club, but there is no house percentage or levy of any kind upon the bets made. I am sure he believes that if he joins, he will after all be able to make his fortune by skilled gaming. I am convinced this will be our ruin. May I call upon you at 12 o'clock today for your further advice?
>
> Yours sincerely,
>
> *Catherine Lawrence*

"Well, I suppose we will have to delay our departure for Sussex, Holmes," I said ruefully. To my surprise, my colleague shook his head.

"The matter is really beyond my remit. I cannot cure all the world's fools. I will leave a note for the lady, advising her that if she cannot dissuade the Marquis unaided, it is time to sever the engagement and seek a more suitable husband. Love is all very well, Watson, but there comes a point where Darwin's laws must be left to take their course."

I was horrified by his callousness.

"I am surprised at you, Holmes! Surely you cannot abandon a client in her hour of need. Especially when you might be considered at least partly responsible for her plight."

Holmes looked at me with a face of thunder.

"*I* responsible, Watson? It was I who saved her fiancé from his earlier folly, after your bungling attempt failed. What action of mine could possibly be held to blame for this latest development?"

"Well, I think you should have warned him off gambling in more unambiguous terms. I know you mentioned some clever exceptions to the rule that gambling is never wise, but quite frankly, you would have done better to tell him that to gamble is always to lose, even if you were fudging the truth a little."

"I will not compromise on the truth, Watson; it is not in my nature. Now, are you ready to leave for Sussex?"

I shook my head stubbornly. "I will follow you on the next train, Holmes. But I will first wait for Miss Lawrence. Thanks to you, I have recently become quite knowledgeable about matters of probability and gambling, and I believe I can advise her myself."

Holmes shrugged his shoulders. "I fear it is a lost cause, Watson. Nevertheless, I respect your dedication. I will hope to see you later in the day."

Catherine Lawrence arrived a little before noon. She did not appear as agitated as I had expected, and she took the news of Holmes's absence calmly. After Mrs. Hudson had brought us tea, she launched somewhat apologetically into her story.

"The fact of the matter is, Doctor, I now feel my letter sent yesterday was a little panicked. The new club has been opened by a German who is presumably of noble blood—he styles himself Baron von Munchausen—and my husband-to-be visited it last night. It really does appear to be an establishment where gentlemen can bet against one another for fun on an equal basis, so that in the long run, winnings and losses will cancel. Perhaps a wise wife should permit her husband a hobby that is likely to be inexpensive, despite my distrust of gambling."

"There are no roulette wheels to be found in this club, then?"

"Nary a one! All gambling is on a private basis, between members seated at tables. The club subsists off bar takings and modest membership fees. Yesterday Lionel was introduced to two games involving coins and dice. He lost a modest sum, but the gentlemen he was playing with, both sons of well-known aristocrats, were obviously bending over backward to show him no unfairness was involved. I am sure it was pure chance that was responsible."

"Pray tell me the details of these games."

"The first involved a sequence of coin tosses. Each of the two players nominates a sequence of three outcomes. For example, the first might pick 'Head-Tail-Head' and the second 'Head-Head-Tail.' A coin is tossed repeatedly, and the results recorded on a chalkboard, until one of the chosen sequences occurs, at which point that player is declared the winner. It seems to me that every sequence is equally likely, so the game really depends on pure chance rather than skill."

I picked up a sheet of paper and made a rough sketch.

"There are eight possible sequences, as I have written. Of course, if order were not important, a sequence of three heads or three tails would be less likely than a mixed sequence. But if the correct order is required, then just as you say, each is

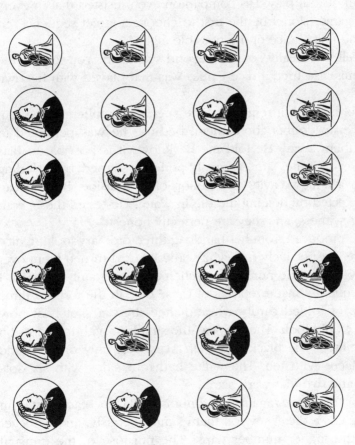

Heads and Tails

equally likely. However, madam, I must tell you that this assumes the coin used is a fair one. Mr. Holmes and I have recently come across a case where a biased coin was used, to devastating effect."

"My fiancé is not such a fool, Doctor. He was invited to use his own coin, and to toss it himself, so there could be no suspicion of foul play. His companion even insisted that in every new game, Lionel be the first to choose a target sequence, so he had first choice of the possible eight."

"Well, that really sounds beyond suspicion. You tell me the Marquis also tried a dice game. Was that played with his own dice?"

"No, because it required three specially numbered dice. But I have no doubts about them, because he was permitted to take them away. He told me he knows it is possible to bias dice, even if they are externally perfectly shaped, by including either a hollow cavity or a dollop of lead inside. But this can be detected by floating the die in water. He tested these, with me as witness, and they are perfectly honest."

She produced from her handbag three ordinary-looking dice colored respectively red, white, and black. I turned them over in my hand. The numbering scheme was certainly unusual. I subsequently made replicas of these dice in the form of three cubes of folded cardboard and then drew a sketch of how each would look. The red was the simplest, with four dots on every face. The black had four faces with only one dot and two faces with ten. The white had four sides with six dots each and two that were blank.

"For this game, again with only two players, each player in turn picks a die. They roll them simultaneously, and whoever gets the higher number wins. The purpose of the unusual numbering scheme is to ensure that there are no draws. The blank faces count as zero."

I calculated in my head. "The red dice will obviously give an average score of four. The white, also four. And the black—

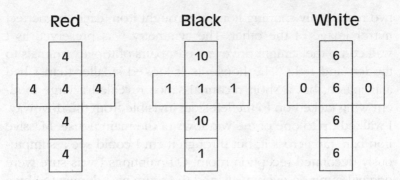

Three Curious Dice

why, that also comes to four. That does seem fair. Nevertheless, I am somehow a little more suspicious of this game. I am not certain the dice are as equal as it appears."

"Lionel felt the same way. But the way the game was played instantly quelled his worries. His host obviously noticed his concern and insisted that in each game, Lionel have first choice. He then took one of the two remaining dice. So if one die is in some way subtly the best, it was Lionel who had the first opportunity to take it."

"Well, that does sound definitive." I rose. "I must leave you now, for Sherlock Holmes requires my assistance in a more serious matter. But I think you should be relieved that your husband-to-be has found a venue where he may exercise his hobby on equal terms with men of his own class. As long as the stakes do not become excessive, and he confines himself to such evidently fair games, you need have no concern. Miss Lawrence, may I offer my confident wishes for a long and happy marriage."

I had assumed that Janus House, Sir James's country residence, would be more than a cottage, but the reality that confronted me was intimidating nonetheless. The gate was guarded by two realistically snarling lions of wrought iron, each the perfect mirror image of the other. The symmetry was preserved as I walked up the straight driveway past pairs of topiary animals to the left and right. The house itself looked smaller than I had anticipated, but a short reconnaissance revealed a large modern wing tacked on to the left side, invisible from the driveway. I walked up to one of the windows of the main house. Massive iron bars ran across it, but through them I could see a sumptuously decorated reception room. Oil paintings I was sure were original hung on every wall, and the room was dominated by a Steinway grand piano and a full-size harp. My observations were interrupted by a hail.

"Well, hello, Watson: I had not expected you for some time yet!" Sherlock Holmes strode up beside me.

"The interview was less difficult than I expected," I explained, "I even hoped I might catch up with you en route."

"Well, you nearly did: I arrived only half an hour ago. Tell me, what do you make of the layout?"

I looked again at the house. I noticed that there were bars on every window, even those on the upper floor. "Sir James seems to have been very security-conscious," I said.

Holmes nodded. "Normally, one would think that a man with such formidable defenses had enemies to fear," he said. "But considering the value of the paintings and other artifacts with which Sir James surrounded himself, his precautions were no more than reasonable. Even though they did not save his life, they should make our task much simpler. Sir James died in his sleep when he was stabbed through the heart at some time between two and four yesterday morning. The doors were barred at sunset, and I am quite satisfied that no intruder could possibly have got in after that time, or got out again after the crime was committed."

"So he was murdered by one of the servants?" I said.

Holmes shook his head. "Sir James was a mistrustful man. He was aware that most art thefts are what the police call inside jobs, involving an accomplice on the premises. The servants all live in the new wing at the back of the house. They withdraw from the main house at 10 o'clock, and Sir James then checks the premises himself and locks the single interconnecting door. No one could possibly get into the main house after that point."

"The crime would seem impossible, then!" I said.

Holmes shook his head.

"No, for he was not alone in the house," he said. "Three of his protégées have been staying there. He was tutoring them in music and deciding which deserved a scholarship that he funded at the Royal Opera."

"Then surely that makes things absurdly simple!" I cried.

"Not so simple as you might think, Watson, no." He glanced at his watch. "But I have made an appointment to interview the first of these young ladies in the drawing room. Pray accompany me and lend your very valuable assistance."

The drawing room was furnished with the same opulence I had already noted. The parquet floor was draped with a tiger-skin rug to which the head was still attached, giving me quite a start as we entered.

The beauty of the young lady who rose to greet us, though, quite eclipsed the splendor of everything else in the room. Long blond hair framed a perfectly oval face. She looked almost more like a Renaissance painter's Aphrodite than a mortal woman. Holmes introduced her as Kitty, and we seated ourselves.

"I understand you have been staying here the past month, so that Sir James could give you intensive tuition upon the piano and the harp, as well as in vocal work," said Holmes.

The girl nodded. "That's right, me and the two other girls," she said. Her crude accent jarred me by its incongruity: her beauty notwithstanding, it was evident that she hailed from a quite humble background. "He wanted also to decide which of us best deserved his operatic scholarship."

"And had he done so yet?"

"He had." A still harsher note came into her voice. "He gave us an examination yesterday in each of our three skills. Then he called us each in turn into his study last night. When my turn came, I was surprised to see that he had before him the day-book. It is a book at the entrance to the music room in which we sign our names, and record the time, whenever we enter or leave the room. We are supposed to put in exactly five hours of practice each day, starting in the early morning, and the purpose of the book is to help us keep track.

"'You have all achieved an irreproachable level in every musical skill,' he said, 'and I have therefore decided to select on the basis of motivation, rather than accomplishment. The scholarship should go to the one who has worked hardest for it. I have decided to adopt a rather old-fashioned criterion. No one supervises you in the mornings, and it is a matter of honor at what time you start your early practices, for no one checks whether the times you write in the day-book are correct. Nevertheless, the order in which your names appear in the day-book is a reliable indicator. It tells me who rose to her duties first, who second, and who last. The scholarship shall go to the earliest riser, like the worm to the proverbial early bird. I had hoped for the best for you, Kitty. But here is the book. You see that Julia was ahead of you on Monday and again on Tuesday. I am sorry, but you know I can only take one of you, and you can see that it cannot logically be yourself.'

"'Do you mean Julia has got it then?' I asked, but he shook his head.

"'I do not wish to sow jealousy among you, when you will all be together for a few days yet,' he said. 'I have asked the winner not to reveal who she is. It is better that way. I am truly sorry, Kitty, but although I am fond of you, I must go with my system. Please do not cry. Ask Julia to come in now, will you?'

"I went out with my head held high, although I cried myself to sleep. I was awakened the next morning by an electric bell the servants could ring to request that the door admitting them to the main house be opened. That was unusual, for Sir James normally rose early and unlocked the door himself. I went downstairs with Alice and Julia—we have separate rooms, but they are adjacent to one another—and we admitted the servants. A little later we heard a scream: Mrs. Hobbs, who had gone to awaken Sir James, had found him dead in his bed. The rest you know." She rose.

"It must have been quite a disappointment to you, not getting the scholarship," I suggested. The girl sniffed.

"I am no worse off now than I was a year ago," she said defiantly. "If service was good enough for my mother, it will be good enough for me. I suppose you would like to see Julia now?"

The girl who shortly entered was of equally breathtaking beauty. Slightly taller than the first, with long dark hair, she seated herself confidently opposite Holmes. Her account was remarkably similar to the first girl's except when it came to the details of the timing.

"He showed me that I had arrived second on Tuesday, after Alice. And then that I had also arrived third on Wednesday, after Alice, who was second. 'As you are after Alice on two days out of three, a bright girl like you will already have deduced that you cannot be first overall,' he said."

"So, Alice did better than Julia, who did better than Kitty, and so she is the lucky one,' I remarked as we waited for the third girl. "That makes her in some sense the odd one out, so we had better pay her the most careful attention."

The third girl was a redhead such as Titian might have dreamed of. I sprang up to assist her to her chair, and tripped over that confounded tiger-skin rug. "By Jove," I could not help stammering, "if you girls are as talented in music as you are striking in looks, you are all three destined for fame, scholarship or no scholarship."

Alice looked at me coolly, and I wished the floor would swallow me up.

"It is not coincidental," she said. "Sir James believed that beauty comes from perfect symmetry of bodily form and that this same symmetry goes with musical talent. We were therefore told that we were selected as much for our looks as for our raw and untrained musical abilities."

Her story continued as I now expected, until the point where it made me sit bolt upright. "Sir James told me that I was behind Kitty both on Monday and on Wednesday, and that obviously I therefore could not be the most dedicated," she said calmly. "He told me that to avoid jealousy, he had directed the girl who had won the scholarship not to reveal the fact, and he expressed his regrets. I went straight back to my room and slept until the servants' bell awoke me."

I could barely contain myself until she left the room.

"Sir James must have changed the book entries at some point," I said indignantly. "The three girls cannot all have lagged the others."

Sherlock Holmes shook his head. "Not so, Watson. The book was found by his bedside. You will see that the order is consistent with what he told each girl." He opened it to the relevant page.

"But that is absurd!"

"No, Watson, what is absurd is the human intuitive tendency to think that a group—be it a group of objects or of people—can always be placed in a unique ranking order by a given criterion. That is true only if we use a linear measure, such as height or weight. If Alice is taller than Kate, and Kate is taller than Julia, then indeed Alice is taller than Julia. But we cannot necessarily say from a more subtle criterion that if Alice is 'better' than Kate, and Kate is better than Julia, then Alice is better than Julia. There is no reason to suppose an unambiguous

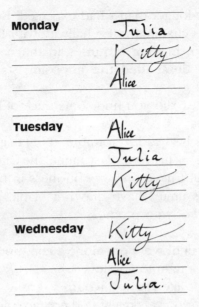

Monday	Julia
	Kitty
	Alice

Tuesday	Alice
	Julia
	Kitty

Wednesday	Kitty
	Alice
	Julia.

The Janus Register

best or worst is defined. In fact, on these lists each girl has come once in first place, once in second, and once in third."

"But what on earth was Sir James playing at? And how does it tie in with the murder?"

"I can make a good guess, Watson. But what I cannot yet say is which of the three girls did it. The situation is too symmetrical. To break the tie, we must seek either a confession or physical evidence. The woman who stabbed her benefactor through the heart in cold blood is not about to break down under questioning, so that leaves only one line of attack. Unfortunately, the local constabulary would not give me access to Sir James's bedroom earlier. But Lestrade has since set them straight, and it is there that I go now. Best if you await me here, Watson: the less disturbance to the scene, the better."

He was upstairs longer than I had expected. By the time he came down, I was chatting in the hallway with Lestrade, who had returned from making inquiries in the nearby village. Sherlock Holmes descended the imposing stairway stone-faced.

"Well, Mr. Holmes, did you find some trace of the intruder?" cried Lestrade.

My friend shook his head. "I found nothing that could reveal which person was in his room last night," he said. "I am very sorry, Lestrade, but I have pressing business in London, and I am afraid that this time I have drawn a blank. By all means keep me informed of your progress."

I held my tongue until we were alone again, upon the station platform awaiting a train to London.

"Holmes, it is not like you to give up so easily," I exclaimed. "Tell me, could your powers really find no trace of the intruder?"

Holmes looked at me thoughtfully. He reached into his waistcoat pocket and carefully removed something.

"On the contrary, Watson, I found this." He held up a strand of hair so blond it was almost invisible.

"So it was Kitty who did it! But why are you shielding her?"

My friend shook his head. "I also found this." He held up a long strand of jet-black hair. "And this." He placed a third length, this one glowing red, across his palm.

"But—good heavens, Holmes—are you telling me all three conspired to murder him?"

"Hardly, Watson. I am sure it was one of the three acting on her own. But murder is not the only reason why a woman enters a man's bedroom, Watson."

"Then you mean—"

"I think your own honesty and chivalry prevent your seeing it, Watson. I am sure that just as Sir James last night convinced each girl she had lost the competition, so in the preceding weeks he misled each girl in turn into thinking he had more than a scholarship for her in mind. Sir James was a very eligible bachelor, Watson, dazzling to a girl of humble background. No doubt each girl was led to think at the time that a sample of passion would seal matters."

"The blackguard! So you decided that the murder was no more than his just deserts?"

"Well, not quite that initially, Watson. I would not say mere seduction was justification for murder. After all, it takes two to consummate a seduction! I expected to find traces of one girl alone, and thus to catch the murderess.

"But the element of poetic justice decided me. Sir James loved symmetry—and too much. The symmetry of perfect beauty can tempt any man, but Sir James sought the vile symmetry of taking advantage of every one of his charges. As a result, the evidence is so symmetrical that at this moment I genuinely do not know which girl did it. And somehow I feel disinclined to dig further. Perhaps Lestrade will solve it, but I have my doubts. Ah, I fancy that is our train on the horizon. Tell me about your interview with Miss Lawrence; it will help pass the journey. Perhaps your day has been more productive than mine."

"Well, truth to tell, Holmes, I had little to do. It turns out the Marquis, rather than reverting to hard gambling, has merely taken to having harmless fun with gentlemen of his own class. The games he is now playing are for small stakes, Holmes, and on terms of absolute equality." I was prepared for his skeptical snort. "I will explain the games to you, and you will see that they are fair."

We found an empty compartment, and I described the dice game to him, sketching again the diagram of how the dice would look unfolded. Holmes looked at it intently and then sighed deeply. "Why, Watson, this is an almost perfect analog of Sir James's scoring system, if we substitute dice color for hair color. You have fallen again for the fallacy of assuming that a set of properties that cannot be reduced to a single numeric value can be used to order objects unambiguously. Though the blunder is not quite so serious when it involves dice rather than young women.

"Consider first the red dice played against the black. Which will win most often?"

I looked. "The red wins 4 times in 6," I said.

"Now play the black against the white."

This was a little harder. "A third of the time, the black throws 10 and wins regardless," I said. "Two-thirds of the time, however, the black shows 1, and then the white wins two-thirds of those cases."

"So the black wins $1/3 + (2/3 \times 1/3) = 5/9$ of the time. The black has the edge over the white."

"I concede: even though the total number of spots on each dice is the same, red wins over black, and black wins over white."

"Now play red against white."

"This should be a sure thing for red, Holmes. . . . Why, no, white wins two-thirds of the time!"

"Quite so, Watson. There is no unambiguous hierarchy. Red is better than black, black is better than white, and white is better than red. It is really no different from the old scissors-paper-stone game of our childhood."

"Pray remind me."

"Hold your hand out of sight, Watson, and at your choice make it flat for paper, or hold index and middle finger apart for scissors, or make a fist for stone. Now bring your hand out as I bring mine . . . now! Ah, I win, for paper wraps stone."

"I remember now: scissors cut paper, paper wraps stone, yet stone blunts scissors."

"The Marquis's dice game is much the same, Watson. It is just that the use of numbers misleads us into thinking that there must be a way to place the dice in sequence from 'best' to 'worst.' I will wager, Watson, that the Marquis was invited to pick the first die every time."

I sighed. "He was indeed, Holmes. And of course if he picked the red die, his opponent chose the white. If he picked white, the opponent chose black. And if he chose black, the opponent chose red. He was on the poor side of the odds in every single roll."

"You also mentioned a coin-tossing game."

I described the rules and drew again my earlier sketch. Holmes frowned thoughtfully and fell into a brown study that lasted until the train slowed as we came to the outskirts of London.

"Why, I have it, Watson! Each sequence is equally likely—but the sequences are not independent in their appearance, because they may potentially overlap."

I looked at him blankly. "You have lost me, Holmes."

"Well, pretend you are the Marquis, and pick H-H-H. I will pick T-H-H."

"Very well. Each sequence is equally likely."

"Yes, but their *relative* occurrence is not independent! Supposing the sequence goes on for a while, and then ends in H-H-H."

"Kind of you: I have won."

"No, Watson. Sketch out any such sequence you like."

I wrote H-T-H-T-H-H-H.

"Now, what is the sequence starting one before H-H-H?"

"It is T-H-H—confound it, that is your choice! Let me try again."

"You will not succeed, Watson. Unless your string actually starts H-H-H, which will occur only one time in eight, it is inevitable that wherever you try to put H-H-H, it will be preceded by T-H-H.

"The subtle thing is that just as with the dice, if you go first, I can always pick a sequence to outdo you."

"*I* will start T-H-H," I said.

"If it pleases you. I will nominate H-T-H. If T-H-H is going to come at any point except the start, there is a 50 percent chance it will be preceded by H-T-H one step earlier. The reverse is not true. It will take a little calculation to prove, Watson, but there is no 'bomb-proof' string: pick as you like, I can always name a string that is more likely to occur first."

I shook my head. "Really, this seems like black magic, Holmes."

"Not so, Watson. But it does go against a false intuition that Nature has hard-wired firmly into our brains: the fallacy of judgement, that people or objects can always be ranked in an order of value, from best to worst, in a sort of beauty contest. Let us be thankful that it is not true."

He sighed deeply. "I was quite looking forward to Mrs. Hudson's fireside scones. But I think we had better call first at the Munchausen club. There is no need to thank me, Watson. After all, you are always willing when it is I who require some assistance."

8

The Execution
of Andrews

I SHOOK MY HEAD AS I surveyed the morning's newspapers.

"I feel almost sorry for the man, Holmes. He has come through such hardship. Still, the odds on his guilt do appear overwhelming. I suppose justice must take its course."

The headlines were indeed unanimous. Their tone ranged from the hysterical "The Rat Must Hang" of the newest tabloid to the more restrained "Guilt May Be Inferred" of *The Times*, but the writers' opinions were equally clear. The very fact that no witnesses survived to testify against Andrews was more than sufficient proof of his guilt. The only difference of opinion concerned whether he should be tried by summary court-martial and shot, or should be turned over to the civil authorities for formal trial at the Old Bailey and hanged. It was scarcely an enviable choice.

The proven facts were extraordinary enough. A month earlier, the Westshire Light Infantry Regiment, stationed at Rangoon, had marched north into Burma upon reports of escalating strife between two tribes in the interior. It was now clear that the reports had been fabricated, designed to lead the soldiers into a trap. The precise subsequent events were unclear, but the final outcome was not: the Westshires had met the terrible fate of being the first British regiment in history to be wiped out to the last man. Or, as we now knew, to the last man save one. For three weeks later, when massive reinforcements rushed from East India had restored Imperial control in the interior, a disheveled and wild-eyed man stumbled into one of the Rangoon guard-posts. His account of the circumstances of his survival was so improbable that the garrison commander had had no hesitation in arresting him as a deserter and shipping him home for trial.

Holmes raised his eyebrows. "You think Andrews's story is incredible, then?"

I hesitated. "Well, when it comes to that, the whole thing is almost incredible, Holmes. It must be the first and only time in military history that a large force has been so comprehensively wiped out that we have no reliable information even as to how it met its fate."

Holmes smiled. "Really? And you with Scots blood on your mother's side. Shame on you, Watson."

"Why, of course, Holmes. I am forgetting the Ninth Legion!"

The story has excited me since I first heard it at my mother's knee. I paced up and down as I talked, keen to tell my friend the details I so well remembered.

"When the Romans conquered England and Wales, they were forced to halt their northward advance when they reached Caledonia. The resistance of the local tribesmen became too fierce for them. After a period of guerrilla warfare and persistent raids from the unconquered territory north of the border, an entire Legion, the Ninth, marched north to complete the conquest of the islands.

"But instead, the Ninth Legion disappeared without a trace! Undoubtedly massacred to the last man, the only time such a thing occurred in the thousand-year history of the great Roman Empire. To this day, no one knows where and how it happened. Caledonia—or Scotland, as it is now called—was never completely conquered.

"The Romans blamed it on the sheer fierceness of the northern tribes, but we know better, Holmes. The feat could not have been achieved without the most meticulous planning and cooperation between the clans, under some unknown general. He realized that if not a single man returned to describe the Ninth Legion's fate, then terror of the unknown—so much more effective than any finite threat!—would keep the Romans at bay thereafter.

"What a battle there must have been! Highland warriors under rigid discipline lying concealed until the word was given. The Roman columns marching on into the trap, unsuspecting, their scouts killed or cleverly decoyed. Then at the word of command, forward! And so—"

Holmes held up his hand in a soothing gesture. I became aware that in the heat of the action, so to speak, I had leaped onto the back of the sofa and was holding my walking stick in the pose of a swordsman, waving it so wildly as to endanger the integrity of our window curtains. Somewhat embarrassed, I descended with care.

"Well, anyway, Holmes, their success proves, in my opinion, that my ancestors were no mere barbarians, but disciplined warriors who formed part of an advanced northern civilization now lost to history."

"I would not dream of disputing the point with you, Watson. But let us return to the present day, with its uncanny echo of those events of two thousand years ago. A force of ten thousand men sets out into hostile territory. Only one comes back. Can we assume that he deserted his colleagues? We must of course have proof beyond reasonable doubt, to convict him of such a grave offense."

"But consider his story, Holmes!" I ticked off points on my fingers. "He says that his regiment was decimated when it was attacked in the middle of the night. The soldiers tried to rally and give some account of themselves, but the situation became hopeless, and the colonel ordered his remaining troops to mount and gallop away. When they regrouped, there were about 1,000 survivors. Andrews was one of the lucky 1 in 10 to escape."

Holmes smiled without mirth. "It is an intriguing word, Watson, *decimate*. Do you know that in Roman times to decimate meant to kill just 1 in 10 of a body of troops? A unit that had been decimated was considered too demoralized to fight, and was not sent back to the battle. But nowadays, *decimate* is increasingly used to mean the loss of not 1 in 10 but 9 in 10, or the great majority. Ah, the relentless march of progress! But in any case, do you consider Andrews's account of the modern decimation of his regiment unbelievable?"

I shook my head. "No, I understand he has given his interrogators a detailed and consistent account, and he could of course have been one of the lucky few. But the odds mount. For he claims the 1,000 survivors were ambushed as they made their way through a mountain pass. Again they fought until the commanding officer (a captain, for this time their colonel was one of the first to fall) sounded the retreat. Again, Andrews insists he took his chances with his colleagues and did not flee until the order *'sauve qui peut'* was given."

I shuddered. "That must be the order a soldier most dreads to hear, Holmes, far more than the command to enter battle. 'Every man for himself, let he who can save himself do so.' The one command that ever permits a soldier to disregard his colleagues and flee, given only in the most extreme and dire of circumstances. In any case, at the end of this debacle, there were just 100 survivors. Andrews claims that he survived against 10-to-1 odds twice in succession by pure luck. Lucky indeed, for he stood only a 1 percent chance."

I counted off on my fingers. "He says the 100 survivors regrouped under the surviving captain and continued south. But soon they were attacked again: they came under overwhelming fire from rebels equipped with modern rifles taken from their own fallen comrades. This time there was no chance to retreat. Andrews was hit by a ricochet that stunned him. He does have a corresponding wound, but of course there is no telling in what circumstances it was really sustained. It may even have been self-inflicted to back up his story. When he came to, Andrews says, it was to find himself the sole survivor in a heap of his fallen comrades: the rebels had evidently left him for dead. Again a 100-to-1 chance!

"It follows that either Andrews is the luckiest man alive, or else he is lying, having survived by dishonorable conduct, deserting his unit. The newspapers are right, Holmes: with odds of 10,000 to 1 against his story being true, however plausible, I am afraid the man must hang."

Holmes pursed his lips. "Odds of 10,000 to 1 against his story being true? There I must take issue with you. Let us start at the beginning. Suppose we have a regiment of 10,000 brave soldiers with no cowards among them. They face an attack so deadly that each individual has a 90 percent chance of death. How many survivors are expected?"

"One thousand, obviously."

"Then they face a second attack, again with each soldier standing a 90 percent chance of dying. How many survivors?"

"One hundred."

"Then they come under fire so devastating that this time each individual stands a 99 percent chance of dying. How many survivors?"

"A single one, Holmes."

"Indeed! And yet we started from the assumption that all 10,000 were brave men. The newspaper editors and yourself have deduced that if a man claims to have survived against long odds, then the chance of his claim being true is of precisely equal improbability, which is a fallacious equation."

"Good lord! You mean that Andrews must be innocent, then?"

Holmes shrugged his shoulders. "Hardly that, Watson. I am merely drawing your attention to how the misuse of statistics can give a spurious authority to a verdict. Andrews may very well be guilty, but it is nonsense to talk of a 99.99 percent certainty, as those editors do. However, the evidence is beyond reach; it is no case for me. The law will have to take its course. I will reserve my attention for criminal matters within my scope." He glanced out of the window. "And talking of criminal matters, here comes a man who is normally an excellent source of such material. I see the good Inspector Lestrade approaching."

The Inspector joined us moments later, looking rather more cheerful than usual.

"You remember the help you gave us in the rigged coin-tossing games, Mr. Holmes? You pointed out that that tossing a fair coin 100 times and getting 63 or more tails had only a 1 percent probability of occurring by chance and was a reasonable proof of a biased coin. I thought you might like to know that, following the successful convictions I obtained, of course"—he puffed his chest out—"it has been adopted as a national guideline and is being applied by forces up and down the country. In pretty much every pub in the land, every coin-tossing game, however traditional and innocent-appearing, has been observed by undercover officers, and we have a rich haul of results. In Rutland alone, for example, thirty rigged games were discovered in a single weekend."

Holmes raised his eyebrows. "That would not surprise me in London, but in a rural country where any strange face in the local pub is regarded with suspicion, it is somewhat astonishing. How many games in total did the officers log?"

Lestrade inspected a densely written sheet of paper. "In Rutland? About three thousand."

"Three thousand?" My colleague half rose. "Lestrade, you must release the Rutland suspects immediately!"

Lestrade looked at him in astonishment. "Release them? But Mr. Holmes, it was on your advice that we arrested them."

My friend clenched his teeth. "Let us for a moment suppose," he said, "that the coin-tossing games played in Rutland are in fact innocent. You observe 3,000 innocent games. I told you that in only 1 percent of such games would you observe 63 or more tails in 100 tosses. But 1 percent of 3,000 is 30— just what your zealous officers found. What you have actually proved, you buffoon, is that the coin tossers of Rutland are probably all honest men!"

Lestrade sat helplessly, mouth opening and closing like a fish. Eventually he managed to rally.

"Not all the results were so innocent," he said weakly. "In Birmingham, two thousand games were observed and forty suspicious cases detected. By your reasoning we should have got only twenty. Those men we shall keep locked up, I think."

Holmes shook his head. He seized a sheet of paper and drew a diagram.

	Normal	Suspicious
Honest	1,960	20
Rigged	0	20

Gambling in Birmingham

"You may infer that of those 40 men, roughly 20 are innocent and 20 guilty," he said. "Taking any one individual, then, there is only a 50 percent chance that he is guilty. Certainly that does not justify keeping him in jail. Those also you will have to release."

Lestrade ran his finger down the sheet. "In Glasgow," he said, "five thousand games were observed, and one thousand were suspicious. What do you make of that?"

Holmes made a second sketch.

	Normal	Suspicious
Honest	4,000	40
Rigged	0	960

Gambling in Glasgow

"Roughly 40 of those men are innocent," he said. "The other 960 are guilty. Any individual is 96 percent likely to be guilty. It would have been preferable to observe them for longer before arresting them, but those suspects I think you may reasonably continue to hold."

The Inspector left us in some confusion, but was clearly relieved to retain some fig leaf of dignity: at least not all of the men whose arrest he had advised would be let go.

"There is something very wrong in this reasoning," I declared after thinking the matter over for a while. "If I had been innocently tossing coins in Rutland, I had a 1 percent chance of false arrest but am now released. But if I had been innocently tossing coins in Glasgow, however, I stood the same chance of false arrest but am still in custody. That cannot be right. I am convicted in the one place and acquitted in the other, just because there are so many con men in Glasgow, although I am not one of them."

Holmes nodded. "Yes, I must face that consequence of the advice I have given Lestrade: some innocent men will suffer," he said. "Guilt is rarely certain, and every judge has on his conscience that a certain percentage of the men he sentenced were probably innocent. But the advice I gave was consistent nonetheless. When deciding the probability of guilt, we should take all the available information into consideration, and that certainly includes the percentage of criminals in the region. One way to look at it is that, in combing a given number of criminals out of society and into prison, it is inevitable that we shall sweep up with them a smaller but proportionate number of innocent people."

I shuddered. "The imprisonment of innocent people is intolerable to contemplate!" I cried. "Really, Holmes, there are times when your cold detachment repels me."

Holmes looked at me with more sadness than I have ever seen in his face. "No system of determining guilt can be infallible," he said. "If you insist that the scales of justice be weighted so that there is zero probability of convicting an innocent, you will never dare to lock up anybody, and all the criminals will also go free. There are two problems: that of false positives, innocent men who are convicted; and that of false negatives, guilty men who go free. We want both percentages to be as low as possible, and most people believe that the threshold of proof should be set so that the first percentage is lower than the second, but errors will inevitably occur in each direction. You cannot escape setting the threshold, Watson—deciding implicitly or explicitly what percentage probability is 'proof beyond reasonable doubt.' I say that even though every society in history has fudged the question of what that percentage should be, preferring not to contemplate the issue too clearly."

"I once did a stint as a prison doctor at Pentonville, Holmes. Conditions were very harsh. The thought that some fraction of those men were in fact innocent is quite horrible."

"Indeed, Watson, and one of several good reasons for tempering punishment with humanity is the inevitability that some of those punished will be innocent."

"You look rather careworn today, Watson," Holmes commented on my return from my rounds later that day.

"Yes, I have had serious matters to occupy me. I was dreading my first patient this morning, Holmes. The last time I saw her, I detected several possible symptoms of Rigor's disease. It can be cured if caught early but is otherwise fatal. However, the cure is an operation that is painful, expensive, and dangerous. You would not want to undergo it unnecessarily. I was in a dilemma as to what to advise her. Fortunately, a new kind of test for the disease has recently been invented, involving a patented mix of chemicals that react with the patient's urine. The test has been rigorously evaluated and has proved to be 90 percent accurate." I waved the relevant copy of *The Lancet* at my friend. "I tested her this morning, and the test proved positive, so I have sent her for the operation."

"A good day's work, Watson. I am glad to see you keep up with new developments in your field."

"Unfortunately, the rest of my day was less satisfying. You see, the test kit I had purchased contained sufficient chemicals to test about a dozen people, but the chemicals remain active for only a day after the seal is broken. You know how I hate to waste things."

"Part of your frugal Scots upbringing, no doubt. So you felt compelled to use up the chemicals?"

"Yes. I had ten other patients to see today, for the usual variety of complaints a GP regularly encounters. Just to put the chemicals to good use, I gave each of these patients a test for Rigor's at the end of the appointment. To my surprise, one came up positive."

"Why did that surprise you? I thought Rigor's disease was quite common."

"It affects 1 person in 100 at one time or another. But the chance that a randomly chosen person is suffering the disease—even in its earliest stages—at a given moment is more like 1 in 500. But the test was positive, so I was compelled to refer the patient for the operation. Poor man, he had come in merely to have a broken finger splinted. My news came as quite a shock to him. Still, I am sure I did the right thing. The test is guaranteed right in 9 cases out of 10, after all!"

Holmes frowned. "Ninety percent accurate. May I see that copy of *The Lancet*, Watson?"

He read the article on the new test. At one point he frowned and pulled down a slender volume from the bookshelf above him. A few minutes later he looked up.

"Watson, I have just decided that the statement 'This test is so many percent accurate' is just about the most misleading thing you can say in the English language!"

"Whatever do you mean, Holmes? I thought you were all in favor of exact percentages, rather than vague intuitive notions, when assessing probabilities."

"And so I am, Watson. But any statement about probabilities is meaningless unless it is put in context. Probability of what, and in what circumstances? To assess the usefulness of a test, whether we are testing for illness or criminal guilt, we must take into account not one but three probabilities.

"First, what is the chance of a false positive? What percentage of healthy, or innocent, people will wrongly be tagged as ill, or guilty, by the test?

"Second, what is the chance of a false negative? What percentage of ill, or guilty, people will wrongly be tagged as healthy, or innocent?

"Third, what is the actual incidence of illness, or criminality, among the relevant population?

"In general, these three figures will all be different. Your experience with the Rigor's test is a perfect illustration. What it actually says here, Watson"—he waved *The Lancet* article at me—"is that the test fails to detect the disease in 15 percent of people who actually have it. Thus, it is 85 percent accurate as regards false negatives. On the other hand, 5 percent of people who do *not* have the disease *do* cause the chemicals to change color. So the accuracy as regards false positives is 95 percent."

"Well, the average of 85 and 95 is 90, Holmes, so for those of us who are not rigorous mathematicians, calling the thing 90 percent accurate is surely a reasonable approximation. I accept the correction, though, and feel sorry for the 15 percent of victims the test will overlook. But you have confirmed my earlier decision about the man with the broken finger. If the rate of false positives is only 5 percent, then there is a 95 percent chance that he does have the disease, and so I must send him for the operation."

Holmes shook his head. "Wrong, Watson! Believable, but utterly wrong! You have forgotten to take into account the third factor: only about 1 person in 500 has the disease at any given time. Thus, the third relevant percentage is 0.2 percent. Let us take a sample of 10,000 people at random from the general population and give them all your new test. Only 20 will actually have the disease. Of those, 17 will test positive and 3 negative. On the other hand, of the 9,980 healthy people, 499 will test positive. All these figures are typical rather than exact, of course.

"So, Watson, you have heroically tested 10,000 patients and summoned 516 back to tell them their tests were positive. How many of those will really have the disease? Just 17. The chance that a random person who tests positive really has the disease is nowhere near 90 percent, as I think you believed, but around 3 percent." Holmes returned to the table and drew a diagram.

"Good Lord, Holmes, I see it now. I must call round on Mr. Peters—the man who came in with the broken finger—first thing tomorrow and set his anxieties at rest." I hesitated. "But

	Ok	Suspicious
Healthy	9,481	499
Sick	3	17

Rigor's Disease

what about the lady I saw first? In her case, it was not just the chemical test: she definitely had other symptoms that suggested Rigor's disease, though I was far from certain."

"Well, when you have seen these symptoms and suspected Rigor's disease in the past, Watson, what has your success rate been?"

I hesitated. "Very roughly, Holmes, about half such patients whom I had intuitively suspected of having the disease actually went on to develop it."

"Then the lady comes from a subpopulation that has a 50 percent chance of developing the disease. If you took 10,000 like her, 5,000 would have the disease, of whom 4,250 would test positive. Of the 5,000 who would not be afflicted, 250 would test positive. So there is a 94 percent chance she has the disease."

I nodded. "That is sufficient to refer her for the operation. I was beginning to think that giving her the chemical test had been entirely pointless."

"Not so, Watson. But you have illustrated the difference between what is described, in police language, as going on a fishing expedition, and seeking evidence against suspects targeted for good reason. You have illustrated the inadvisability of the former and the good sense of the latter."

I sat back in my chair and loosened my waistcoat. I should have been relieved, but I could not help feeling rather foolish to have so greatly misestimated probabilities that now seemed rather obvious to me.

"I feel very stupid. But I must say, Holmes, you have seen through potentially misleading data with your usual admirable clarity," I admitted.

He smiled. "You should thank not me, Watson, but a remarkable mathematician called Bayes." He held up the book he had referred to earlier to show me the name on the spine. "He was the first to correctly analyze such cases—and not because his predecessors were all stupid. Bayes's theorem sets out formally the criteria for calculating probability ratios such as those we have been encountering today."

"I will be sure to credit him if I write up today's events. If you show me it, perhaps I should reproduce his formula to illustrate the point."

Holmes turned the book toward me to reveal, I must say, a rather intimidating piece of algebra.

"I would not advise it, Watson. I have heard it said that every equation appearing in a popular book halves its sales: your fear of algebra is not unique. I confidently predict that if this formula appears in all its glory, your sales will be decimated—and in the modern sense of the word! No, you should confine yourself to illustration by example. Those window-frame-shaped diagrams I have been drawing for you summarize Bayes's approach exactly."

He leaned forward intently.

"In any case, it is not so much for his formula as for the clarity of his thought on vexing questions of probability that Bayes is remembered—his philosophy that when calculating probabilities, you must take the global picture into account. False positives, false negatives, the global frequency of what you seek, and any biases that may arise from the limitations of your data and your observational vantage point. The simple-sounding message has endless ramifications. Mathematicians frequently hold conferences on what is called Bayesianism, and there are so many subtle issues that I would bet they will still be running those conferences a hundred years from now."

"Well, Holmes, I am sure his work has many ramifications. But I have learned just a single lesson: to be mistrustful of percentage claims that sound good, but are not, when people are trying to sell me things!"

Holmes smiled. "There are positive sides, Watson. Let us apply Bayesianism to the assessment of your good self. You reckoned that your intuitive assessment of Rigor's disease is confirmed about 50 percent of the time, which you do not seem to consider very good. From that, we can say something about your rate of false positives and false negatives.

"Your rate of false positives must be astonishingly low. For even if you correctly picked up on every actual case of the disease—100 percent on avoiding false negatives—it is so rare that you must be wrongly identifying only 1 in 500 healthy people as having it—a false positive rate of just 0.2 percent."

"Oh, I doubt I am as good as that," I said, blushing.

"Well, if your ability to spot actual cases is only 50 percent, your false positives must be even lower at 0.1 percent. The truth is probably somewhere in between. If you are as good at spotting actual cases as the chemical test, 85 percent, then your false positive rate is less than a sixth of a percent, some 30 times better than the chemical test. You are a better doctor than you give yourself credit for, Watson!

"By the way, there is a further application of Bayes's theorem I have made today. You remember the case of Andrews? It is now definite that he will be tried at the Old Bailey."

I could not see how Bayesianism could help the unfortunate private, but I nodded.

"I thought you had decided not to involve yourself in that case, Holmes."

"I have changed my mind. The young man's sister called on me, to plead persuasively for my help."

"Surely she is not exactly an unbiased witness."

"Well, the man's previous military record was certainly excellent. There are no a priori reasons for assuming him a coward. Anyway, call me soft-hearted if you like, but I have agreed to be called as an expert witness in the boy's defense. The trial starts next week. I will let you know when I find out what day I am to be called."

Despite my long acquaintance with Holmes, I have only once or twice been in the vast cathedral of a building that is the Old Bailey. As we made our way down broad, uncarpeted corridors toward the Central Criminal Court, it was difficult not to feel intimidated by the echoing hugeness of the place. It must be quite terrifying to come there as the accused. At the entrance to the court, an usher separated us, and I made my way into the public gallery just in time to see Holmes sworn in at the witness stand. The court was dominated by the vast judicial bench, at which Lord Donaldson presided in wig and robes. I knew that if the verdict was death, the judge would don a black cap over the wig—actually a merciful gesture so that the returning defendant would know his fate at once and be spared the agony of further uncertainty. Andrews himself was a small, white-faced man, sitting alone in the huge wooden dock that has housed so many of the notorious criminals of our era.

The prosecuting barrister was Major French, a man also famous for his role in harsh military courts-martial. However, he made the great mistake of underestimating my colleague.

"Mr. Holmes, I believe you have come to plead there are no surviving witnesses against Private Andrews. But it is precisely because he is a deserter that he is the sole survivor! The evidence against him is statistical alone, but odds of 10,000 to 1 constitute proof well beyond reasonable doubt."

"I quite agree," replied Holmes quietly. "Statistical evidence alone should be enough to convict or acquit. May I beg the court's indulgence while I educate it in the relevant statistics?"

To my astonishment, he launched into an account of my experiences with the test for Rigor's disease. People in the gallery, and even the judge, turned to look at me as he spoke. Eventually he was cut short by the judge's gavel.

"We have indulged you, Mr. Holmes, but you really must explain why this is in any way relevant to the case before us."

"Certainly, your honor. I would like to ask a question of Major French. Major, I believe you have served in action with two of the Queen's most famous regiments. Have you seen any men executed for cowardice?"

"Yes. Five, after the battle of Madras."

"A battle fought under extreme conditions, was it not? With about 5,000 soldiers on the British side, I believe. But the heat of battle uncovered only 5 cowards among them: 1 in 1,000."

Major French drew himself up. "Certainly I do not believe the British Army is overstocked with cowards. But on your figure of 1 per thousand, there would have been 10 such as Andrews in his expeditionary force. Your line of argument is getting you nowhere."

He folded his arms with a satisfied look, but Holmes continued undaunted.

"Let us consider the question of false positives and false negatives. To keep the numbers simple and whole, let us imagine an army of 100,000 that is reduced to 10 or so men: the same survival factor as that observed in Andrews's regiment. It initially contains 100 cowards and 99,900 brave men. Let us further suppose that a coward who deserts has a chance of survival 100 times better than a man who stands and fights. That is certainly an overestimate, for desertion carries its own risks. It is not for nothing that when British soldiers go into battle, their military police form a line immediately behind them."

The major reddened but did not challenge Holmes as he proceeded. "Suppose they face such odds, then, that a brave man stands 1 chance in 10,000 of survival, and a deserter 1 in 100. Thus, 10 brave men will survive the battle, but only 1 coward. An individual survivor has only 1 chance in 10 of being a deserter."

Holmes strode forward and rather theatrically unveiled a large sheet of artist's paper on which he had drawn a diagram, its shape by now quite familiar to me.

	Die	Survive
Brave	99,890	10
Cowards	99	1

Survivors and Cowards

"Now, Andrews's regiment consisted of only 10,000 and had only 1 survivor, but the ratio of the odds is the same. It is 10 to 1, Major, that Andrews is no more a coward than you are!"

The acquittal followed the same afternoon.

"Do you really think Andrews was as brave as his fellows?" I asked as we walked back to Baker Street.

Holmes shrugged. "I must confess to a certain manipulation, knowing that the major would not admit of the possibility of a

British regiment containing more than a very few potential deserters," he said. "There are really infinite degrees of bravery and cowardice. There are heroes, there are brave men, there are ordinary men, and so on. Most men have a breaking point at which they turn to flee, and whether they are court-martialed depends in practice on the extremity of the circumstances."

Holmes looked off into the distance. "We cannot know the precise degree of cowardice or bravery shown by Andrews. But I thought of his solo trek back to the coast, crawling through hundreds of miles of enemy territory, only to find his own countrymen treating him worse than the enemy would have done. Justice must be tempered with mercy, Watson. Unlike bacteria, human beings have feelings, and if your profession is rightly ruthless toward microbes, mine must always be biased toward avoiding false positives, even if it produces a few false negatives. Andrews and his family have suffered enough. My conscience is at ease with today's outcome."

9

Three Cases of Relative Honor

THE SHARP DOUBLE-KNOCK AT the door had the ring of military authority. My Service days are well behind me, but the stance of the smart young officer who stood on the threshold brought back my parade-ground days so vividly that I almost found myself saluting him.

"Is Mr. Sherlock Holmes at home?"

"No, sir—that is to say, no, he is currently at the French Embassy, but he is expected back in half an hour or so."

"I will call back. Would you be so good as to tell him that Captain Henderson of the Seventh Royal Hussars wishes urgently to see him?"

"May I ask what it pertains to?"

The man's bearing became even more rigid, if possible. "I am shortly to face court-martial. I stand accused of killing twenty men."

"And you believe Sherlock Holmes may be able to prove your innocence?"

"No, I wish him to discover whether I am guilty or not. For I have no idea myself!"

And with that he gave a stiff bow, clicked his heels in an almost Prussian manner, and was gone.

I shook my head. I have studied amnesia: true cases are very rare indeed. Yet those who stand accused so often ask the jury to believe a remarkable lapse of memory! Well, if Henderson was in fact guilty, and had deluded himself that involving Sherlock Holmes in his case would be to his advantage, he was likely to receive his just deserts.

Holmes returned before the half-hour was up. He flung himself into his armchair with a sigh.

"How went it at the Embassy?" I asked.

He buried his head in his hands and groaned. "No luck at all, Watson. And it is hardly surprising. Consider the obstacles I faced!" He counted on his fingers. "First, when the break-in is discovered, the ambassador and his staff make a bungling attempt at detective work. The next day Sûreté detectives arrive from Paris and make a second investigation, which is also fruitless. Only then does the ambassador decide to waive the building's diplomatic immunity and allow Scotland Yard in. Lestrade's men almost pull the building apart in their efforts to convince the French of their zeal to solve the case. Finally, they think to summon me. Really, Watson, it was like trying to find the trail of a mouse in a field where three separate herds of elephants had subsequently run amok!"

I sighed. "And yet Lestrade is convinced it was only a common burglary. Really, Holmes, you should have told them that the matter was beneath your dignity."

"I am certain Lestrade is right, Watson. He even has two suspects in custody. But the French are convinced the matter was more sinister. They believe the burglary was mere cover for a more sinister violation of diplomatic immunity: the copying of secret documents. I became involved because Mycroft begged me. He believes that if the case cannot not be cleared up promptly, a major deterioration in relations between the Great Powers is imminent. More I cannot tell even you."

"And *was* it a mere burglary, Holmes?"

"Undoubtedly, Watson. Lestrade has the right men: a pair too dim even to realize that the building was an Embassy and the hue and cry that would result. They are called Ludd and Johnson. The first is a former boxer who, punch-drunk, can no longer fight; the second is a street vendor unable to make a living at his honest trade. A fine irony, that two of the stupidest men in London should cause such a furor."

"Well, if they are so dull, surely one of them will slip up in questioning before long, and suspicion will turn to certainty?"

My friend shook his head. "Do not underestimate the power of stupidity, Watson! A stupid man can maintain a stubborn silence where a clever one might be provoked into a slip. They are both hardened criminals, and only a foolish optimist would expect an admission from either."

At that point there came again the sharp military knock on our door, and I realized I had forgotten to warn Holmes of our visitor. Before I could say anything, however, Holmes had sprung up and opened the door himself.

"Mr. Sherlock Holmes? I am Captain Henderson of the Seventh Royal Hussars. I believe that you recently intervened in the court-martial of a soldier called Andrews, very possibly saving him from execution. I also am shortly to face a court-martial."

My friend waved him to a chair. Mindful that Henderson claimed to have killed twenty men, I picked up the poker and pretended to poke at the fireplace with it, before laying it down in easy reach beside my chair.

Holmes coughed. "Please excuse my colleague, Captain. He often feels the need to check the status of our fireplace, in high summer. What is the crime of which you are accused?"

"I am charged with negligently causing the deaths of twenty men under my command."

I perceived that I might have been a trifle hasty with the poker.

"And are you guilty?"

"I do not know."

Before Holmes could reply, I spoke, I am afraid, rather hotly. "Captain, I am a medical man, and I can say in all modesty that I am an expert in amnesia. I am convinced that true cases of complete amnesia are rare—very rare indeed. If you claim such a convenient gap in your memory, I am afraid we will treat your story with great skepticism."

Henderson looked calmly at me. "I am not claiming amnesia. I remember the relevant events with crystal clarity. The deaths certainly occurred, and as a result of my decision. But was my decision negligent—as is now being charged—or was I merely the victim of bad luck? That is the question."

Holmes leaned forward, waving me into silence. "Pray tell us the story."

Something in his manner seemed to reassure our visitor, who relaxed somewhat. But at that moment there came a loud crash from the street. A common enough occurrence, yet the effect on Henderson was extraordinary. He sprang to his feet, then flung himself on the floor behind the sofa, hands over his head, quivering all over.

Holmes leapt to the window. "It is only a brewer's cart, striking the curb."

He assisted our visitor back to his chair, while I splashed some brandy into a tumbler. For the first time I felt some sympathy for the young captain. In Afghanistan, I had seen men shot for cowardice whose behavior suggested nervous disorder rather than any ordinary lack of courage. Soldiers who had killed too often at close quarters sometimes suffered "trigger finger": their hands became paralyzed, and they would not or could not continue shooting fellow human beings. And others exhibited "shell-shock" after surviving an explosion nearby. Perhaps someday these would be considered medical conditions, rather than mere lack of courage.

Henderson gulped the brandy and shortly was able to speak again. "My regiment has just returned from a tour of duty in East India. You are doubtless aware that sporadic trouble continues there, led by a tribe known as the Mauras. At the point in question, I was leading a platoon on a reconnaissance patrol that had lasted several days. We were trying to verify rumors that the insurgents, until then lightly armed, were being supplied with heavy weapons from lands to the north. We were crossing a marshy plain, known locally as the Baar Valley, that is overlooked by high mountains to the north."

He pulled a folded piece of paper from his pocket, and spread it to reveal a roughly sketched map.

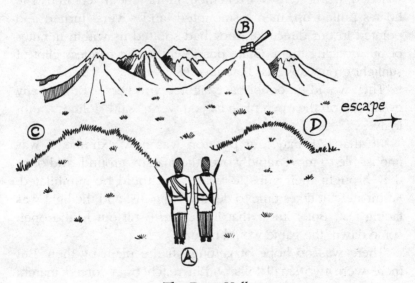

The Baar Valley

"It was early evening, and I was about to call a halt for the day, when one of my corporals shouted that he could see movement on the clifftops to the north. We set up our field telescope, and when I brought it to bear my heart sank. I saw a long rebel column, with horses pulling sizable artillery pieces: the rumors were true. But more alarming still, I could see signs of a much larger gun that was being towed in several parts. There had been whispers of this monster, said to be capable of firing exploding rounds of a hundredweight each: it was referred to as Big Bertha."

"Why, how extraordinary!" I said. "My service was in Afghanistan, a decade ago, yet rumors of a gun of that name circulated there."

Holmes smiled. "The nickname Big Bertha has been used many times, in different theaters of war, to refer to the largest gun known in those parts, real or imaginary," he said.

"Well, this one was no figment of my imagination. You may think things could hardly have got any worse, but as I watched, I saw signs of excitement in the enemy column. The horses pulled up, men dismounted, and several turned and pointed in our direction. They had spotted us within minutes of our noticing them. Quite possibly it was a careless glint of sunlight on my spyglass that gave us away."

"They would no doubt have sighted you very shortly in any event, out on the open plain as you were," said Holmes soothingly.

"Be that is it may, our position was now extreme. It was impossible to move rapidly over the marshy ground, and once they brought their guns to bear, we should be annihilated. Fortunately, it takes time to deploy big guns, and the light was fading fast. But I knew that if we were still out in the open come dawn, the game would be up.

"There was no hope of getting off the plain by then. But there were two small hills within reach by a forced march. They were not large enough for the locals to have given them names—I have labeled them simply C and D on my map, as you can see—but either would provide enough cover to keep us safe from the enemy's conventional artillery until night fell again. The problem, however, was Big Bertha. A single one of her shells fired at high elevation could plummet down behind either hill and would be sufficient to blast any force hiding there into eternity."

"A hopeless situation, then," I said.

"Not quite! For the great disadvantage of these very big guns is that they cannot be fired at all frequently. For hours after each shot, the barrel is too hot to handle, and then repositioning the gun and sighting it in on a new target is a most laborious process. That is why these huge pieces are really of more propaganda value than actual practical use on the battlefield.

"By the following dawn, we could be concealed behind either hill, and the enemy would not know which one. During the day, there would be time to fire one shot only from Big Bertha. The enemy would have to take a chance on which hill to aim at. The question was which hill we should pick. We had a little time to make our choice, for obviously we should not give away our intentions by starting to move until it was fully dark. I debated the point with my master sergeant: I am not the kind of officer who is too arrogant to think his subordinates may have useful opinions."

"Surely you might as well have tossed a coin," I said. "Evidently one hill was pretty much as good as the other. Debate is one thing, dithering is another. One can end up like the proverbial donkey, starving to death midway between two identical bales of hay because it cannot choose which to go toward."

Henderson shook his head. "The symmetry was not so perfect as you suppose, Doctor. You see, hill D was quite close to the edge of the plain. If we chose hill D, and Big Bertha did not get us the following day, we were home free—we would be able to walk to safety during the night. If on the other hand we chose hill C, then even if we survived Big Bertha, our safety would not be guaranteed. The situation might have changed in our favor or against us: for example, the weather might close in, making long-range artillery sighting impossible, or reinforcements might arrive. The future is ever uncertain. But to a first approximation, if we chose hill C and Big Bertha did not get us, our chance of survival would still only be 50 percent. Whereas if we chose hill D and escaped Big Bertha, our chance of survival would be 100 percent."

"Then hill D is obviously the better choice," I said.

Henderson looked coldly at me. "But the tribesmen above us, knowing the local terrain, were equally aware that hill D was from our point of view the better bet. So it seemed to me that they would be more likely to aim Big Bertha at hill D."

"So then, hill C should have been right for you," I conceded.

"Doctor, the first lesson of military strategy is never to underestimate your enemy. I thought that the tribesmen might well anticipate our reasoning and fire at hill C after all."

"Then you could have done a double-bluff," I said. "If you thought the tribesmen were devious to that degree, hill D should have brought safety the following day and escape the night after."

"And if the tribesmen anticipated that also, would you then try for a triple-bluff?" said Holmes with a smile. "Do you not see, Watson, you are entering an infinite regress. This reasoning will get us nowhere."

Henderson shook his head despairingly. "I carried on a similar debate with my master sergeant on that fateful evening," he said. "He was for hill C, and I was for D. We could reach no sensible conclusion. The thing that particularly troubled me was that the big gun we could see being assembled was undoubtedly of Prussian manufacture, supplied to the rebels by the Germans. Now, where German artillery goes, German military advisers tend to accompany it. I could be sure the decision as to which hill to fire on would be made by a Prussian officer. They are trained far more rigorously in tactics and strategy than we British: I knew his logic would be impeccable in deciding which hill to aim at. If only I could duplicate that logic, and anticipate his decision, we would survive! But the more we argued, the more uncertain the decision seemed.

"In the end I almost resorted to tossing a coin as you suggested, Doctor, but it seemed cowardly to dodge making a decision in that way. I was responsible for the lives of my men; I could hardly delegate so serious a choice to the fall of a coin. I overrode my sergeant, and we marched to hill D and were well concealed behind it by dawn.

"By midday the next day there had still been no sign from the enemy. But seconds after midday a tremendous thunder-clap smote our ears. Big Bertha had fired! We waited in an agony of suspense for what seemed like hours, though in real-ity it can have been no more than seconds. Then there came to our ears what was at first a faint whistling sound but grew louder rapidly, so rapidly. . . . "

He buried his head in his hands, his shoulders shaking. Holmes splashed a further liberal measure of brandy into our guest's glass and forced it into his hands. But this time Hen-derson waved it aside. In a few seconds he was himself again.

"I came to my senses in what seemed an eerie silence," he continued. "In fact I had been deafened by the blast, and it was days before my hearing returned to normal. I was sur-rounded by bodies, freshly turned earth mercifully concealing the details of their features. The enemy had outguessed us, and the shell must have come down right in our midst. By some fluke of the blast pattern, I alone had survived. At first I wished I had died also, but eventually I recalled that I still had a duty to return to my unit and report, and that night I made my way to safety."

"And was your story not believed?" asked Holmes.

"Oh, yes, I was believed. No one suggested I was guilty of desertion. But I have been charged with the lesser, yet still very grave, offense of dereliction of duty. By my recklessness in ignoring my sergeant's advice—and although he was nomi-nally my junior, he was an older and more experienced soldier than myself—I caused the deaths of all my men. My court-martial takes place tomorrow. If I am found guilty, at best I can expect to be reduced to the ranks, and at worst drummed out of the army in disgrace."

He looked Holmes in the eye. "Now, sir, I am no man to shirk responsibility for my decisions. If I was indeed reckless, I will deem my punishment fair. But the question squirrels round and round in my brain, almost driving me to the point of madness: Was I indeed reckless? Or was I merely the victim of bad luck? For I am no martyr, and if it was only bad luck, then the court-martial is a mere sham of finding a scapegoat for defeat. I cannot decide the matter: I am a man of action, with no claim to be a philosopher or mathematician. But it occurred to me that what was hopelessly baffling to me might be simplicity itself to Mr. Sherlock Holmes."

Holmes rose and clapped the man on the shoulder.

"I will take your case. When does the trial take place? Thursday, at the Old Admiralty buildings by Trafalgar Square. Very well: you may put down my name as an expert witness for the defense. Have confidence; I shall be at your side."

But when I returned from showing Henderson out, it was to find Holmes with a troubled expression.

"Come, Holmes!" I said. "In recent months, your ability to solve these problems of logic and probability has impressed me greatly. Surely this two-choice question does not present you with any difficulty?"

"On the contrary, Watson. The problems we have confronted so far have involved fixed probabilities set by the laws of Nature. Now Nature is subtle, but never malicious, and the calculations have been relatively straightforward. But here we are dealing with an intelligent opponent who is actively trying to evade our analysis and confound the logic of our choice. The rules of this game are much harder."

"It reminds me of a much more innocuous kind of shell game that I played as a child," I said thoughtfully. Holmes cocked an eyebrow inquisitively.

"One child conceals a sweet under either of two seashells, and the second child tries to guess which," I explained. "I am afraid I was rather bad at it. I was usually the hider. I do not believe any cheating took place, but whichever shell I put the sweet under—whether I kept to the same choice as on the previous turn or whether I changed it—I generally seemed to get outguessed."

Holmes smiled. "If you have a fault, Watson, it is perhaps that you are a bit predictable," he said. "My advice comes a little late, but I think you should have resorted to tossing a coin to choose your shell."

"It seemed to me, as to Henderson, that it would be rather lazy to thus dodge the effort of making my own decision."

"Not at all, Watson. We saw in the case of Madam Zelda how bad the unaided human mind is at generating random numbers. It simply cannot avoid falling into a pattern of some kind, be it more or less obvious. Sometimes you need an external source of randomness if you are to avoid predictability. Paradoxically, delegating a decision to the fall of a coin can be the most logical thing to do!

"But this is a more complex case. What should Henderson have done? Follow his more experienced sergeant's hunch? Toss a coin after all? I am blessed if I know. No, pray do not interrupt. This is quite a three-pipe problem, and I shall be most grateful if you will leave me undisturbed to consider it."

When the lunch bell sounded Holmes did not appear, and I went back upstairs to summon him. I found him staring into space with a darkly brooding look, drumming his fingers nervously upon the arm of his chair. Before him was a sheet of paper with a simple diagram.

H

		C	D
	C	0	100
M	D	50	0

C or D?

"I see you have made some progress, although I cannot quite make it out," I said, indicating the sheet.

He looked at me as though from far away. "Scarcely, Watson. That is merely a chart of the four possible outcomes. The columns represent Henderson's choices and the rows the choices of the Mauras. Henderson can choose C or D to hide behind in darkness; meanwhile, the Mauras are lining their cannon up on C or D. The numbers represent the platoon's percentage chances of survival. If both parties choose C, or if both choose D, it is zero: the platoon is annihilated. If Henderson chooses D and the Mauras C, the platoon gets clean away. If Henderson chooses C and the Mauras D, it is fifty-fifty."

He shook his head ruefully. "It is an absurdly simple diagram. Yet somehow the answer to the conundrum must lie within it."

"Well, at least you have done some work. I think you can knock off for lunch with a clear conscience. After all, Henderson is not on trial for his life."

"In a sense, he may be, Watson. Proud young men like that have been known to fall on their sword in such circumstances, rather than live out the shame. But I confess the problem grips me for another reason also. Henderson's dilemma seemed to ring a faint bell with me: I had some feeling of déjà vu. Yet I have never known a case like it. Eventually the source occurred to me."

He pointed to the diagram. "Instead of H for Henderson and M for Mauras, think H for Holmes and M for Moriarty. Does anything suggest itself?"

I stared at the diagram in bafflement.

"Do you remember a day some years ago when we fled London aboard a train for Dover? At that point I knew my only proper refuge from Moriarty was to leave the country; I was planning to take the ferry to France."

"But Moriarty chartered a special train and came after you. He would have killed you had he caught you. How could I forget!"

"Fortunately, the express I was aboard made an intermediate stop, at Canterbury. I had the choice of disembarking there, still in the country, but at least alive for the moment— provided that Moriarty chose Dover as his destination. If Moriarty and I both chose the same destination, I was certain to die. If I went to Dover and he to Canterbury, I was safe. If I went to Canterbury and he to Dover, I had a fifty-fifty chance." He pointed to the diagram. "For H read Holmes, for M read Moriarty, for C read Canterbury, for D read Dover. *That is the very same dilemma I faced that day*. Now, Moriarty was a very clever man, and a mathematician to boot. I knew he would use the most immaculate logic in choosing whether to direct his 'special' to Canterbury or to Dover. Do you know how I chose my destination, Watson? By the fall of a coin!"

"I never saw you toss a coin on our journey."

"I did not need to. I simply put my hand in my pocket and felt the outer side of the first coin I grasped. I have very sensitive fingertips: I can tell the face of a coin by touch alone. It was tails, so we went to Canterbury. I needed you to have faith in my decision, so I did not mention how arbitrary my choice was."

I looked at him in some concern. "Those were indeed desperate days, Holmes. But it is all in the past now. History records that you chose Canterbury, and Moriarty Dover, and Moriarty is long dead at the bottom of the Reichenbach Falls. So put the matter aside, and come and have some dinner, I beg you. Irregularity of mealtimes is not good for the constitution."

"No, thank you, Watson. I find that just as Henderson feels a burning need to know whether bad judgment or bad luck caused his men's deaths, I have an equal need to know: Was my survival that day owing to good judgment—or merely good luck? I will not rest until the matter is clear in my mind."

Having seen him in such moods before, I left him to brood. But when I returned from my afternoon rounds it was to find the air thick with tobacco smoke and Holmes still sitting rigid with a gaunt, white face, doubtless reliving those awful last days of Moriarty's reign. I decided that the time had come to risk his ire.

"Holmes, this brooding is not good for you. If you will not listen to me as a friend, then hear me in my capacity as your medical advisor. When I come across a case that baffles me, I am not ashamed to consult a specialist in the field, nor should you be. You are a detective, after all, not a mathematician."

Holmes turned an expressionless face toward me. "And whom do you suggest I might consult, Watson?"

"Well, you could try Mycroft. He must dabble in figures all day, in his capacity as special advisor to the Government."

Holmes frowned. I knew he disliked having to appeal to his brother for help—their relationship was not without a certain sibling rivalry!—and I pressed on hastily: "After all, he did not hesitate to ask for your help with this French Embassy case. He would merely be returning the favor."

Holmes looked at me angrily, and I stiffened. Then suddenly he tossed his head back and laughed. "Oh, very well, Watson; your common sense will not be denied! I will take a stroll down to his club and see if he has anything to offer."

He returned sooner than I expected, carrying a large, leather-bound volume prominently stamped "Diogenes Club Library."

"Mycroft was never a man to waste words, Watson. But he advised me to read this."

He proceeded to do so, turning the pages with lightning speed, as was his custom. Doubtless Mycroft had lent him some book on military strategy, I mused, very likely Clausewitz's classic work. Shortly, Holmes took a sheet of paper and commenced to scribble figures on it. At length he put the book down, and I was able to read the title on the spine: *Game Theory*.

"Really, Holmes, I am ashamed of you!" I expostulated. "I thought you were working on Henderson's case, and here you are idling away your time with a book that tells you how to play better bridge, or some such."

He looked at me mildly. "This is not a book about card games."

"Well, backgammon, then."

"I think you have been misled by the book's title, Watson. The mathematical analysis of such pastimes is called *games* theory. *Game* theory, by contrast, concerns itself with more serious things. In particular, the tactics that should be adopted in competitive situations, where you can gain only at the expense of your opponent: such situations are called zero-sum games. They are commonly encountered in business and in war."

"I beg your pardon, Holmes. And has it enlightened you?"

"Very much so. The problem was that Henderson had to aim for one of two hills. But apparently he could not make his choice by logic alone, because it was essential that his actions not be predicted by the enemy—that they be to some extent random.

"My earlier diagram was misleading, because it implied that he had only two choices. Yet, as I suspected, the answer did in a sense lie within it. I had simply written too few rows and columns, ignoring intermediate possibilities. What Henderson really had was a choice of *strategies*, in the sense that he had a choice of what *probability* to assign to heading for hill D rather than hill C.

"That is summarized on this new diagram. Let us suppose Henderson makes his choice by rolling a die. Before throwing it, he decides which faces stand for hill D as opposed to hill C. The seven columns represent Henderson's possible strategies. He can assign a probability ranging from zero out of six to six out of six to heading for hill D, depending on how many faces of the die he designates D.

H to D with probability:

		6/6	5/6	4/6	3/6	2/6	1/6	0/6
	6/6	0	8	17	25	33	42	50
	5/6	17	21	25	29	33	37	42
M to D	4/6	33	33	33	33	33	33	33
with	3/6	50	46	42	38	33	29	25
probability:	2/6	67	58	50	42	33	25	17
	1/6	83	71	58	46	33	21	8
	0/6	100	83	67	50	33	17	0

H's survival probability (%)
C or D?

"But of course the Mauras will be trying to outguess him. They also should randomize their choice to make it unpredictable, and they also have a choice of strategies. Let us suppose that they also throw a die to make their decision. Then the rows of the table represent the Mauras' possible strategies: the probability that they aim Big Bertha at D rather than C."

I blinked at the complex array of figures.

"Henderson wants to choose a column that maximizes his chance of survival. But the Mauras will pick the row that minimizes it. Hence arises the concept of the *minimax,* beloved of game theorists. We must look for the column in which the *lowest* value is as *high* as possible."

I ran my fingers across the columns. "In column 1, the minimum is 0. In column 2, it is 8. In column 3, 17. In column 4, 25. In column 5, 33. In column 6—ah, we are back down to 17 again. And then to 0 in column 7. So Henderson should pick column 5, which means he should head for hill D with probability one-third and for hill C with probability two-thirds."

"Now tell me the Mauras' strategy. The Mauras want to minimize the maximum probability of his survival, of course."

I looked down the rows. "The Mauras should select the third row down and aim for hill D with probability two-thirds and for hill C with probability one-third."

"Congratulations, Watson! That is what each should do. Far from intuitively obvious, was it not?"

I found myself hesitating. "I am not sure it is intuitively obvious to me even now, Holmes. It seems awfully complicated. Would it not come to the same thing if they each tossed a coin to decide?"

"Look at the table, Watson. That is equivalent to choosing the 3/6 row and column. That gives a survival probability of 38 percent, which suits Henderson, but does not suit the Mauras. So they are unlikely to adopt such a scheme."

I nodded. But then a more profound thought struck me.

"If your opponent is a good game theorist, Holmes, then actually it does not matter whether you follow the table. Because if the Mauras play row 3, as according to you they should, then it is irrelevant what Henderson does, because every probability in that row is the same: 33 percent. Similarly if Henderson chooses column 5, as you say he should have done, then the Mauras' strategy makes no difference. It is only if *both* deviate from game theory that there is any change."

I thought a moment and then went on in excitement. "It is a little like the lookout paradox. Every ship at sea, even in mid-Atlantic, is required to keep a continuous watch so as to avoid collision with other ships. But in fact if you trust every *other* ship to obey the law and itself keep a good lookout, then it does not matter whether you do so yourself or not, for the other ships will avoid you."

"Until another captain thinks the same way," interjected Holmes. "No, Watson, you might get away with a lazy strategy decision a single time, or even several times, but once your enemy realizes your choices are insufficiently randomized, you are done for. If in the next Great War British officers toss coins to make decisions when they should use more subtle calculations, and the enemy realizes this, then the enemy will have the upper hand.

"An example strikes me that is entirely too close for comfort. Recall again that chase eastward from London, where Moriarty and I had to choose between Canterbury and Dover? Moriarty knew we were both clever men, and he may have assumed I knew my game theory. Accordingly, he would have thrown a die, assigning the numbers to give a two-thirds probability of his going to Dover, which is indeed what happened. But in reality I was ignorant of game theory and tossed a coin. In effect, I chose column 4.

"Now, it is possible that Moriarty guessed I did not know game theory and would resort to tossing a coin. In which case Dover was definitely the best bet for him. If Moriarty felt sure I was ignorant of game theory, then he went to Dover not as a result of a dice throw, but as a result of choosing row 1: Dover with 100 percent certainty. Which gave me a probability of survival—the minimax for that column—of just 25 percent: 1 in 4, as opposed to 1 in 3."

"Well, it does not matter now, Holmes. As it turned out, you went to Canterbury, and survived; Moriarty is dead, and can never tell us on what basis he chose Dover. All else is moot."

Holmes looked at me without seeming to see me, his gaze focused somewhere beyond infinity. "Is it, Watson? Do you remember the many-worlds view of reality, endorsed by Challenger and many other clever physicists, that arises out of quantum theory? That logic indicates we actually inhabit a multiverse in which countless possible realities play themselves out."*

I shuddered. "I remember that the logic seemed unassailable, but it still makes me dizzy to think about it."

"In that case, the original Sherlock Holmes who tossed a coin on the way to Canterbury gave rise to a huge (but not infinite) number of subsequent versions. Call that number a zillion if all had survived. If I had rolled a die as I should have

* See Colin Bruce, *The Einstein Paradox, And Other Science Mysteries Solved by Sherlock Holmes* (Perseus Books, 1998).

done, a third of a zillion would be alive now. As it is, there are only a quarter of a zillion. One-twelfth of those other versions of myself were killed by my stupidity."

I gazed into the fireplace for some time, musing like Holmes on philosophical realities almost impossible to grasp. Eventually the chiming of the clock recalled me to the here and now.

"I suppose it will go harshly with Henderson," I said. "He certainly did not select the right tactics."

"On the contrary, my dear fellow; I anticipate that my testimony will exonerate him."

"What! But how?"

"The Empire expects its officers to be very courageous and reasonably intelligent, but it does not expect the impossible. If I appear on the witness stand and testify that in a similar situation, I, Sherlock Holmes, was unable to make the decision correctly, I can hardly imagine they will find him guilty."

He sighed and gestured to the morning paper. The headline, referring to the French Embassy case, read "Sherlock Holmes Baffled by London's Stupidest Burglars."

"I cannot honorably do otherwise, but the press will no doubt have a field day at my expense again. Humiliations rarely come singly. Clearly, the gods feel that I need keeping in my place."

By the end of the week, both of Sherlock Holmes's predictions had come true. Henderson was indeed acquitted. But the Sunday papers were remorseless at my friend's expense. Particularly cruel was a cartoon on the front page of *The Messenger* that showed Holmes wearing a dunce's cap, sitting at the back of a class of police detectives. Fortunately, I reached the breakfast table ahead of him and hid *The Messenger* under a pile of other papers.

Scarcely had Holmes sat down, however, when the door opened again and my friend's expression changed to one of surprise. Turning, I saw Mycroft standing in the doorway.

"Good morning, Sherlock," he boomed. "I fear you have been having rather a rough time of it of late. The newspapers are absurdly unfair, are they not!" And with a superficially sympathetic smile, he pulled from his coat pocket the very cartoon I had sought to conceal from my friend.

Holmes read the page poker-faced, but I could tell he was affected. Nevertheless, he gestured for his brother to join us at the table and poured him some coffee. "I doubt that you made the journey here just to be amused at my expense, Mycroft. What can we do for you at this early Sunday hour? Has the French Embassy crisis grown still more serious?"

"No, it is blowing over by itself, as these diplomatic flaps tend to do. But it has led to something of interest. I came here primarily to thank you, Sherlock."

My friend looked blank. "To thank me? For what?"

"Well, your failure to solve the French Embassy case by your industrious manual efforts set me to thinking whether there might not be more cerebral ways to prove criminal guilt. And then when your query the other day reminded me of game theory, the solution was not far behind. Congratulate me, my dear fellow: I have invented a foolproof way to extract full confessions in any case involving two or more suspects working together!"

My friend raised his eyebrows. "That is indeed remarkable. I would be fascinated to hear the details, Mycroft."

His brother responded by pulling from his pocket a sheet of paper bearing a diagram that looked rather familiar. "Take the crooks Johnson and Ludd. They are both in custody at the moment, due to be tried for several minor burglaries they committed together. They will certainly be found guilty of those and be sentenced to a month each. The French Embassy was a more serious matter: for that, they could expect twelve months each, except that we have no proof.

"You are aware that judges give a reduction in a sentence for what is called 'turning Queen's evidence'—that is, for making a full and voluntary confession of all crimes and implicating any accomplices."

"How large a reduction can such a prisoner expect?" I asked.

"It depends very much on the circumstances. Normally a confession merits a sentence reduction of 20 percent or so. But if the prisoner implicates an accomplice who might otherwise have gone unpunished for the crime, he can get his own sentence suspended: he walks free immediately.

"Now, I arranged for Ludd and Johnson to be held at separate police stations so that there could be no possibility of communication between them. Johnson is at Vine Street and Ludd at Bow Street. Then I went to see Johnson, who appears marginally the brighter of the two. He, like his colleague, has so far refused to admit anything. I pointed out that he would certainly get a month for the minor burglaries, and he obviously understands that. I also pointed out that if the Embassy burglary could be proved against him, he would get not one month but twelve. Then I explained the offer I had obtained a judge's permission to make. If he turned Queen's evidence, and Ludd did not, he himself would go free, and Ludd would get twelve months. I was quite honest with him, and admitted that if both he and Ludd turned Queen's evidence, neither would go free. But they would get a slight reduction in their sentences: from twelve down to ten months each, in acknowledgment of their confessions."

"Well, I would not confess, if I were Johnson," I said. "Obviously the best thing for both is not to confess, and to wait out the month."

Mycroft nodded. "I am not surprised to hear you say that, Doctor," he said. "But I think Sherlock is a little ahead of you. For I explained my next point to Johnson most carefully. From Johnson's point of view, there were two possible scenarios. Ludd might confess and implicate him, or he might not. But either way, Johnson would be better off confessing! Look at the diagram. If Ludd has not spilled the beans, then Johnson gets off scot-free by confessing, whereas if he does not confess, he serves a month. On the other hand, if Ludd does spill the beans, then Johnson can still reduce his sentence from twelve months to ten by confessing. So whatever Ludd does, it is logically better for Johnson to confess.

Ludd

	Mute	Confess
Mute	J = 1 L = 1	J = 12 L = 0
Confess	J = 0 L = 12	J = 10 L = 10

Johnson

Forced Confession?

"I left Johnson to think it over and went to Bow Street to see Ludd. He is even dimmer than his colleague, but I managed to explain things to him in the same terms." Mycroft simpered. "It is really quite diabolical in its cleverness, is it not? Both would be ten times better off if neither confessed—getting one month each rather than ten—but each individual is compelled to own up by simple logic! The following day, I went back to see each again, confident of full confessions."

Sherlock Holmes smiled. "And with what result?" he asked.

Mycroft slammed his fist down on the table in frustration, spilling my coffee into its saucer. "None whatever!" he cried. "Neither would confess. The problem is evidently that they are so monumentally dense that they are unable to follow simple logic. It would be in the interests of each to confess, regardless of what the other does, but in practice neither one will."

"So why have you come to me?" asked my colleague.

"Because, Sherlock, even though your intellect sometimes lacks a certain cutting edge when compared to mine," said Mycroft, tactlessly tapping the newspaper cartoon, "you have in one respect an advantage over me. You are good at explaining things to persons of limited intelligence. Why, sometimes you even manage to explain quite abstruse matters to Watson here."

I was reminded why, despite his great intellect, I could never be as fond of Mycroft as of his brother!

Sherlock nodded. "I will be glad to help you, Mycroft. There is indeed a point that needs putting across. Could you meet Watson and me at Bow Street in an hour or so? There is someone there who, I believe, can explain it even better than I."

"I am almost impressed, Holmes, by the proverbial honor among thieves that Ludd and Johnson display," I commented as we walked to our rendezvous. "I can only suppose that, despite their low criminality, they have a kind of ethic that allows them to have implicit trust in each other, negating Mycroft's scheme."

Sherlock Holmes snorted. He appeared to be in high good humor. "I think not, Watson. It is remarkable how even an intellect such as Mycroft's can become so focused on a particular problem that he forgets it is inextricably part of a larger picture—and that it makes sense only in that context. Mycroft is not alone. From the book he lent me, it is evident that some of the greatest mathematicians in the world similarly tried to construct game theory in an abstract void, remote from reality, and caught on to their error only rather late."

We arrived a few minutes early. I had thought we would be meeting one of the bright young inspectors there, but Holmes merely introduced himself at the front desk and made arrangements for the prisoner Ludd to be transferred from his cell to an interview room. At the appointed hour, Mycroft appeared.

"Your error, Mycroft, and that of other theorists, has been to consider the classic Prisoner's Dilemma as a problem in isolation," said my friend severely as we walked down the police-station corridor. Mycroft turned an annoyed face toward him.

"If Johnson and Ludd knew that after serving their sentences, they were each to be transported to different colonies, so that they would never see one another again, and if they were also certain that no news of their past behavior here in Britain would ever catch up with them, then they would indeed have a dilemma. But of course that is a quite unrealistic and naive scenario. Game theorists frequently cannot understand why parties to a 'deal' do not cheat, when it is apparently in their interests to do so. Yet deals are not once-in-a-lifetime occurrences. A businessman must engage in regular transactions, often with the same partners, to make a living. If he cheats someone, that person will not deal with him again. Moreover, word that he is not to be trusted may get around. Hence, the businessman deals honestly, not out of some ethical sense but out of self-interest. The same applies to a criminal. A burglar who peaches on his mates must face consequences, including a reluctance of those or other persons to partner him in future ventures. The word gets around."

Mycroft frowned thoughtfully. "You are implying," he said carefully, "that a mathematical calculation that takes into account multiple iterations of a situation will arrive at different optimal choices from those that arise in a single isolated interaction."

I must confess I blinked in some confusion, trying to understand this remark. Sherlock Holmes flung open the door to the interview room.

"Well put, Mycroft," he said. "But here is someone who can explain the point even better. Ludd, I believe you have already met Mr. Holmes senior?"

The burly occupant of the room glared at us suspiciously.

"Ludd," said Holmes slowly and patiently, "I am not trying to trick you. You are in no jeopardy, and nothing you say now can be held against you. I would like you to explain to my brother why you do not want to peach on your mate."

Ludd gazed at the expectant Mycroft with the air of one regarding a half-wit, and spoke deliberately: "Cos he'd bash me head in when he got out."

"It is a regrettable sentiment, Watson, *Schadenfreude*. And yet very much part of the human condition," Sherlock Holmes said unexpectedly as we strolled back to Baker Street, this time taking the longer but more pleasant route through Regent's Park.

"I beg your pardon?"

"It is a German word that describes taking pleasure in other people's misfortune. Our hypocritical English language has no word for it, but others do. A Chinese proverb comes also to mind. Loosely translated: 'There is nothing nearly so satisfying as seeing an old friend, dearly beloved and respected, slip in the mud and sit undignifiedly on his behind in the sight of the entire village.'"

He pulled the newspaper cartoon with the dunce's cap from his pocket and regarded it ruefully. "I believe the British public has a certain respect for me, yet it will have enjoyed today's story all the more for it. And similarly, I yield to no one in my respect for mathematicians in general, and Mycroft in particular, yet how wonderful to watch one of London's stupidest men lecturing its cleverest! It has softened the blow to my own pride somewhat.

"But not altogether erased it. Talking of learning from foreign cultures, I see our route takes us past *Le Canard Enchanté*, where I happen to know they have just imported a superb new wine. Let us go and drown our sorrows, Watson. If Mrs. Hudson loses patience and feeds our dinner to the cats, then so be it."

10

The Case of the Poor Observer

"I SOMETIMES WISH, WATSON, THAT you were a little less zealous in writing up our adventures for popular consumption!" Sherlock Holmes held up a sheet of closely written paper. "What do you make of this? I have been getting a number like it since you described our adventures with the Marquis of Whitewater and Madam Zelda in *The Strand* magazine."

I scanned the page. A Major Blenkinsop, writing from Warwick, claimed that he had been dealt a royal flush at poker twice in one evening. He wanted to know whether my colleague felt this could be chance alone, or whether he should suspect the dealer of cheating.

"I should say suspect the dealer, Holmes. The odds against such an occurrence must be phenomenal!"

"Indeed, Watson. In round numbers, the chance of a royal flush hand is about 1 in 650,000."

"Tell him to have the dealer drummed out of the club forthwith!"

"Not so fast, Watson. I must confess I would be very suspicious at such an occurrence happening to me personally. But if you consider how many dedicated poker players there are in this country, the total number of hands dealt every year must be in the billions. Suppose a typical player is dealt 65 hands in one evening. The chance of two royal flushes is about 1 in 100 million. If the country contains 4 million poker players playing on average two evenings per week, the major's experience will occur some four times every year. Thanks to your literary efforts, I am evidently now viewed countrywide as a reporting-center for such occurrences. So I am not in the least surprised to receive regular letters about freakish card hands. Indeed, it would be more surprising if no such flukes were reported.

"It gives an interesting insight into how, when assessing improbability, you must take the observer's circumstances into account. From Major Blenkinsop's point of view, it is a remarkable thing to get two royal flushes in one night. From my point of view, though, his letter is quite unastonishing. After all, there are millions of Major Blenkinsops out there, and those of them who are dealt unexceptional poker hands do not write to inform me of the fact!"

I felt quite relieved by Holmes's explanation: I could picture the conscientious major rather vividly, dithering in the dilemma of causing a scandal locally, perhaps implicating a personal friend.

"Then there is no reason for him to be suspicious," I said gladly.

Sherlock Holmes shook his head.

"That is assuming far too much. Cheating at poker is not very rare, whereas naturally occurring royal flushes are. Without more data, I could not even venture a guess as to the probity of Major Blenkinsop's colleague. I will write back to say so."

I returned from my rounds later that day to be intercepted in the hallway by Mrs. Hudson.

"Begging your pardon, Doctor, but a gentleman called who is most insistent on seeing yourself or Mr. Holmes."

"I see. Would you happen to recall his name?"

"Yes, Doctor. He is Mr. Rolleman. Mr. Barnum Rolleman."

I smiled indulgently. "I fear you must be mistaken, Mrs. Hudson. Mr. Rolleman cannot have been here."

She drew herself up indignantly. "Why are you so certain I should be mistaken, Dr. Watson?"

"Because I pronounced him dead upon the floor of his hotel room a year ago!"

At this point a voice spoke from behind me:

"But I am here, Doctor."

The accent was American, the voice unmistakably Rolleman's, but there was an eerie quaver in the tone that I would never have expected of that gentleman. I do not mind admitting that I gave quite a jump. But when I turned, it was a young man rather than an elderly ghost who stood doffing his hat in the further corner of the hallway.

"I am Barnum Rolleman the second. It is sometimes the custom in American business families to name the son after the father. I am sure you did what you could to help him. And now I in turn need Mr. Holmes's advice most badly."

I led him up the stairs, doing my best to ignore Mrs. Hudson's rather smug look, and sat him in our visitors' chair.

"Mr. Holmes is out at the moment. Please tell me your trouble; perhaps I can be of assistance."

He nodded. "On my father's death, I inherited leadership of his very large business empire. I have been trying to run it according to my own style, but I very much fear it is escaping my control. I had previously run the New York part of the business quite effectively, I think. But running a worldwide empire with ten thousand employees is quite different from managing a thousand or so in departments I can personally visit and keep in touch with. I have been increasingly out of my depth."

I had a distinct feeling of déjà vu. His predicament sounded very similar to that of my cousin James. I had recently run into

James, and had learned that his previously troubled cab business was not only surviving, but prospering more than it had ever done under his father. I clearly remembered the advice my colleague had given him, and I do enjoy an opportunity to show Holmes that there is the occasional case I can solve without his assistance. "I think I can help you," I said. "Here in Britain we understand such things as the penny-wise, pound-foolish fallacy, the prior investment fallacy, and the cab driver's fallacy."

Rolleman looked at me with scorn. "I learned about those while I was still in short trousers," he snapped.

I was therefore rather relieved when the door behind me opened and a familiar voice spoke: "I think, Watson, that the problems of a business empire spanning two continents and dozens of different activities are likely to be of a different scale from those of a cab firm!"

Holmes took a seat and nodded to our visitor as he filled his favorite pipe. "Pray continue your account. You managed the New York departments successfully, you feel?"

"I do, if I say it myself."

"To manage even a thousand employees, you must have mastered the art of delegation. Is managing ten thousand so very much more daunting?"

Rolleman nodded. "I delegated, but to men I knew personally. I could inspect and verify things with my own eyes. I have identified the new problem: it is not that I am incapable of making decisions on a larger scale, it is that I do not have the *information* on which to base them. I do not mean that I do not have *data*—I could bury myself under endless reams of figures if I wanted to—but I cannot distill it into a meaningful overview. With only written reports and figures to go on, I cannot tell what is really happening."

"Perhaps you are too hard on yourself," I said. "After all, it takes some time to learn the ropes in any new job. Surely your father appointed competent managers for the various divisions of the company. Cannot you leave things to them, be patient, feel your way, until you have a better grasp of things?"

Rolleman shook his head. "No, Doctor, it is more urgent than that. Any large organization has the parasites of dishonesty and corruption ever nibbling at its extremities, like barnacles upon a ship's hull. If such cases are nipped in the bud, the losses are small. But let there be any hint that central control is faltering, that local managers are getting away with scams, and problems mushroom: in no time the ship is sinking. If I cannot get a grip quickly, my father's empire will be done for."

He looked up appealingly. "I realize that this is a little different from your usual sort of case, Mr. Holmes. But it is not for my own sake alone that I am asking. I have two goals for the business. One is that it will grow and remain profitable, true; but I also want to see it treat its employees better. At present the managers earn huge salaries, the ordinary workers very little. I want to restore the balance somewhat."

Holmes nodded thoughtfully. "I think I can help. Give me an example of a problem you suspect."

Rolleman waved his hands helplessly. "I hardly know where to start. Each case in itself seems trivial. Take the fishing at the country club. I am suspicious, but the problem might be no more than a figment of my imagination."

"Pray go on."

"It was a new scheme of my father's, still under construction when he died. He reasoned that the businessmen of New York would pay well to enjoy the traditional countryside pursuits of hunting, fishing, and shooting. Accordingly, he was building a huge club-hotel in upstate New Jersey. One of the facilities was a great shallow pond stocked with a thousand carp, five hundred of which were golden-scaled and easy to spot. Even the poorest fisherman would get a reward for his efforts before long! But to provide a challenge for the more sporting, the other five hundred were black, much harder to see. The black kind are more expensive, but my father did not stint when he felt a thing needed to be done properly.

"I have my doubts about the contractor who stocked the pond. If I had been present to supervise personally, I would have checked each delivery of these fish as it was made, for that is the only reasonable opportunity to count them. But by the time I got up there, the fish had all been delivered. I could have been cheated in either of two ways. First, there may be fewer than a thousand fish in all. And second, perhaps more than half are of the cheaper golden type. Indeed I took a boat out on the lake, and I counted several gold for every black one I spotted."

Holmes shook his head. "But that is an example of biased observation. The gold fish were easier to spot in the first place. What you should have done—"

"I am not quite a fool, Mr. Holmes. I understand the technique of sampling. I sent two men out to fish by night, casting blind so there would be no bias. They caught a hundred fish, and roughly half of them were indeed black. So that suspicion was unfounded."

"Was that not rather a waste of fish?"

"Not at all. They threw the fish back unharmed, after cutting a tiny mark on the fin of each one, so that they would know they were not catching the same fish over and over. But now we come to the harder problem. How to know that there are really a thousand fish in the lake? That problem cannot be solved by sampling. I cannot conceive of any way to count them reliably, without catching them all or draining the whole lake."

Holmes smiled. "But it can be solved by sampling, Mr. Rolleman, and the task is half done!"

Our visitor gazed at Holmes doubtfully.

"You caught a hundred fish and marked them by cutting their fins. The marked fish have had ample time to mingle with the rest again. Now you must send your men out on a second night to catch a second batch of a hundred.

"If there are indeed a thousand fish in the pond, then 10 percent have their fins cut, and so about ten of the fish they catch should be so marked. If, on the other hand, you have been cheated and there are only, say, five hundred, then 20 percent have been marked, and so about twenty of the second catch will be found marked."

"Why, that is so obvious now you point it out, Mr. Holmes!"

"Obvious in retrospect, perhaps. It took zoologists a long time to think of it. But it has become the most reliable method of calculating the numbers of wild populations—estimating how many fish there are in a whole ocean, for instance."

Our visitor sat back and took a deep breath. "I shall wire instructions to do as you suggest. But I fear we now come to more intractable problems."

Rolleman paused to collect his thoughts and then resumed. "I decided that I should spend some months traveling and seeing for myself what was going on in the business worldwide. But there was a problem. I wanted to do my observing incognito, because when people know the boss is watching, they may act quite differently from their normal practice. The presence of the observer can affect what is observed."

"Yes, indeed—it is a little like the observer problem in quantum physics," said Holmes. But at Rolleman's blank stare, he waved him to continue.

"Anywhere I went in America, I would be instantly identifiable. But on this side of the Atlantic, no one would recognize me. My father owned two significant businesses in London: the famous shop Barnum's, and also the new motor-omnibus company."

I started. The motorized buses that he referred to came down Baker Street at frequent intervals, but I had had no idea that they were a venture of Rolleman's. On occasion I had also shopped at Barnum's. The huge store was famous for stocking items of every size, from the tiniest to the largest, in its numerous departments. But its unique feature was its self-service system. Rather than being waited on by an assistant, customers were free to wander about the shop, placing any item that took their fancy in specially provided baskets. When they descended to the ground floor, they lined up at cash registers to pay for the items. It sounds down-market, but I have found the system remarkably convenient: the attention of snooty salesmen and saleswomen can be quite intimidating to an unpretentious man like myself.

"The new omnibus service was making less profit than hoped. That was not too surprising, for it was an innovative and speculative venture. More worrying was a dramatic decline in profits at Barnum's. I decided that before making myself known to either business, I would travel the buses and shop at Barnum's incognito, and find out what was really going on. I was so determined that none should anticipate me that I booked passage across the Atlantic under a false name. I used the same false name to register at an unpretentious hotel.

"The following day, I set out to test the omnibus service on the route that runs from Golders Green to Hackney, passing your rooms here, I believe." He smiled. "It is perhaps the first time a multimillionaire has traveled on such a conveyance. I dressed in humble street clothes, with a workman's cap pulled well down. I waited at the eastward-going Marble Arch stop.

"I was a little disappointed that I had to wait over 20 minutes before the bus came. They are supposed to leave Golders Green every 20 minutes. But obviously traffic introduces a random element; perhaps I had merely been unlucky. I tried again that afternoon. This time I had to wait half an hour! I became obsessive, taking the bus to and fro until I had a valid statistical sample.

"It is obvious that if the buses really did come regularly every 20 minutes, one should on average have only to wait 10 minutes. Even if there is random variation, as long as the total number of buses per day is constant, the average waiting time should presumably be unaffected, longer gaps being balanced by shorter ones. In fact, my average waiting time was a good half hour—three times longer than it should have been.

"My first thought was that the manager was simply running fewer buses than he should. Saving himself money, and no doubt lining his own pocket with the difference, at the expense of hundreds of ordinary people having to wait in the rain. To prove it, I went up to observe the Golders Green terminus. To my surprise, buses were in fact leaving there every 20 minutes, just as they should be.

"Then I thought that perhaps some buses were leaving but were not actually traveling the route—perhaps they were doubling back to the garage. I found a place to spy on the Marble Arch stop where I had spent so much time waiting, a window table in a nearby coffee house. What I observed was truly remarkable."

He paused dramatically. "When I observed from the coffee house, I found that buses did indeed come every 20 minutes on average. There was variation, to be sure. Sometimes two or even three buses came in quick succession; sometimes there was a gap of more than half an hour. But on average, three buses per hour, incontestably." He smiled without humor. "But whenever I paid the waitress and went out to the stop myself, I had on average to wait nearly half an hour! There is only one possible solution to the conundrum."

He banged his fist on the table. "Gentlemen, I am being deliberately defied. I am myself being spied upon, and the London employees are deliberately thumbing their noses at me! When I am seen to come out and wait at a bus stop, the word goes out to divert or delay the next bus, just to spite me.

"It is not only the buses. I have also been to Barnum's on several occasions. On descending to the ground floor with my purchases, I carefully count the number of people in the line before each till; then I always join the shortest.

"There is then an infuriating wait. It is the one disadvantage of this method of shopping. Most customers pay for their purchases promptly and efficiently, but sometimes there are delays. A price ticket is not clear, the customer cannot find her purse, the correct change is disputed—that kind of thing.

"Gentlemen, even though I always join the shortest queue, it promptly becomes be the slowest moving, and I am generally delayed longer than customers at a corresponding position in the other queues! It is quite maddening. Once or twice could be bad luck, but such a consistent pattern can only be a conspiracy by the staff to inconvenience me."

He broke off to gaze in astonishment at Sherlock Holmes. My friend's shoulders had been shaking for some time, and he now lay back in his chair and roared with laughter. Rolleman sprang to his feet, white-faced.

"By all that's holy, if you think I will stand to be laughed at openly by you as well as defied by my staff, you are very much mistaken! Mr. Holmes, you can rest assured that—"

But by this point my friend had regained control of himself. He leaned forward with a conciliatory gesture.

"It was not at you I was laughing, sir, but at the Universe: the way the laws of nature can seem positively to conspire against us. I believe I will be able to set your mind at rest, but you must give me a day or so to research things. You mentioned having a mound of data at your disposal? Shall we say tomorrow evening at seven, and if you bring those books of figures with you, we shall tackle the problems properly."

The following day, I was kept out late by a case involving a young mother running a worryingly high fever. I made it back to Baker Street just before seven, to find Rolleman not yet present but the floor of our living room covered by a large tarpaulin with small objects, perhaps the size of pistols, poking up beneath it. Just beyond the edge, a straight line had been chalked on the carpet.

"Pray do not tread on it, Watson. This arrangement represents some hours of painstaking work."

"I can see that, Holmes. I would advise you to clear it up before Mrs. Hudson sees it. At a guess you have been reconstructing a murder scene, and the chalk line is a bullet trajectory?"

"Completely wrong, Watson. But here comes Mr. Rolleman. I shall save the explanation for both of you. . . . Good evening, sir. I wonder if I you could assist me in a little demonstration.

Be so good as to take this handful of pebbles and walk along the line, depositing them wherever you please."

Rolleman snorted but did as he was asked. Holmes inspected the result and smiled in satisfaction. "Very good! You will agree that I did not influence you in any way. I can now tell you that the line represents a 2-hour period at the bus stop opposite our rooms in Baker Street. Your placement of the pebbles represents the random arrival of persons at the bus stop. Let us see how long they have to wait for a bus. First, we will suppose that in a perfect world, buses arrive once every 20 minutes."

He rolled back the tarpaulin to reveal a second chalk line on which those toy painted omnibuses you can buy for a penny had been placed at regular intervals. That chalk line was designated line B.

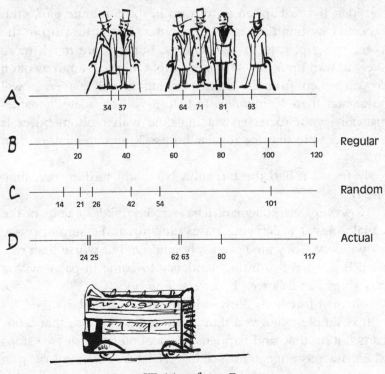

Waiting for a Bus

"How long have these people had to wait?" He took a tape measure from his pocket and bent down. "The first one, about 6 inches—an inch represents a minute on the scale I have drawn. The second 3 inches, the third 16 . . . yes, the average comes out to 10 minutes, near enough, as expected. Now let us suppose that the buses arrive at random moments." He rolled the tarpaulin back further to reveal line C and handed Rolleman the tape measure. Rolleman paced rapidly along the line, muttering to himself.

"The average is 18 minutes. Now, how can that be, Mr. Holmes? The number of buses on the line is the same as before, so the average distance between them must be the same as before."

"Indeed it is. The shortest interval between buses is 5 minutes and the longest 47, but the average is the same. But now consider. If you happen to turn up in the 5-minute gap, your expected waiting time is 2 1/2 minutes. If you turn up in the 47-minute gap, it is 23 1/2 minutes. But you are much more likely to turn up in the 47-minute gap! Only one-tenth as often do you benefit from the shorter waiting time. The average will be greater than 10 minutes. In fact, if you calculate it mathematically, your expected waiting time with random buses is exactly double that for regular buses: 20 minutes rather than 10."

My friend rolled the tarpaulin back still farther, revealing line D.

"However, Mr. Rolleman, I have today made a note of the actual times at which your buses drew up at the stop opposite my window. They are not merely random, but worse than random! Buses have a distinct tendency to come in pairs, with a very short gap followed by a very long one."

Rolleman frowned. "Now, why the devil should that be?"

"It is simple when you think it through. Suppose that a bus starts out on time and happens to encounter one or two stops where no passengers are waiting. The bus can carry on past those stops at full speed, and thus it begins to catch up with the bus ahead of it. Because it is closer to the bus ahead, there is little time for new passengers to accumulate at each stop. The bus ahead acts as a sort of snowplow, clearing the passengers the second bus would otherwise have to stop for. The second bus goes faster, the gap becomes smaller, and the second bus draws still more ahead of schedule, until it is just behind the first.

"It is what an engineer would call *positive feedback.* By the way, that term is often misused in the business world to mean a good report. But to an engineer, positive feedback is a problem to be avoided, defined as a self-amplifying discrepancy that gets worse over time.

"Note that the same effect works in reverse. If a bus falls a little behind schedule, it will on average find more passengers waiting at each stop. Accordingly, it will make slow progress and fall still farther behind."

As my friend spoke, Rolleman had been measuring the bottom line. "The average waiting time in practice is nearly half an hour. Just as I found at Marble Arch."

Holmes smiled. "It is quite a paradox, Mr. Rolleman, is it not? A stationary observer notes an average interval of 20 minutes between buses in all three cases. But an observer who catches the bus waits for much more than the expected 10 minutes. He is likely to arrive at the stop in a long gap rather than a short one."

He waved Rolleman to a chair. "The afternoon I spent noting bus times. But this morning I spent at Barnum's, where I purchased these charming model buses one at a time. I spent the entire morning in line at one cash register or another."

I coughed to interrupt at this point. "I had been forming some opinions on that, Holmes. It occurred to me, for example, that if you notice the queues on either side of you, both left and right, there is a two-thirds chance that one or the other will be moving faster than yours. And if you crane your neck to watch all the queues, then it is statistically very likely that yours will not be the fastest. Perhaps a sort of psychological envy factor causes you to notice the faster queues more than the slower ones."

Holmes shook his head. "That is by no means the whole explanation. However, the matter turned out to be very much simpler than the bus problem. It was quickly evident to me that not all cash register operators serve at the same speed. Some are quick and efficient, some slow and lackadaisical. And then there is always the trainee, all thumbs and hesitation, a supervisor hovering behind to increase her anxiety.

"Habitual customers at the shop soon realize this. They can spot the fast and slow servers. They join not the shortest queue, but the one in which they reckon they will get to the front most rapidly. This being so, the shortest queue that you see inevitably turns out to be the slowest-moving one. The occasional businessman or mathematician who wanders into the shop is competing against professionals—masters of the art of queuing who positively enjoy shopping. It is no wonder the amateur loses out.

"There may also be a selective-memory factor at work. Suppose one day you are lucky and join a short, fast-moving line. You will barely remember having to wait in line at all. But a day when you are unlucky, and are detained a long time, will make a deeper mark on your memory. You could look at it this way: if you dip into a moment in your past at random, then of those moments you happen to have been standing in a queue, more will be found in the longer delays than in the shorter ones. Hence you gain the impression that you are often unlucky. It is very similar to the bus stop paradox: traveling in time, you are more likely to arrive in a long wait than in a short one."

Rolleman sank into a chair with a sigh. He brushed back his hair with one hand. I hastened forward with the humidor; it seemed an appropriate moment to offer him a cigar. He bit the end off and looked my friend in the eye.

"Well, I must say you have already been worth whatever fee you may charge me, Mr. Holmes. You have indeed put my mind at rest. I must find some way to make sure my buses arrive at regular intervals."

"I am afraid I have nothing practicable to suggest there."

"Well, I cannot expect you to do all my work for me! I will think about that in my own time."

He opened a bulging briefcase he had brought with him and pulled from it a thin folder of correspondence, followed by a vast pile of accounting ledgers.

"You asked me to bring my data for your perusal. Here it is. The problem I have is not so much with profits. With the exception of Barnum's department store, whose recent accounts I have here"—he placed a batch of green-bound ledgers to one side—"things are quite acceptable. I am more concerned about the adoption of my social policies. I want to ensure that employees at all levels earn fair wages and also that the customers are fairly treated. If possible, I would like to be a little generous to those who cannot afford the best. For example, while I would like to see my buses as full as possible for a high load factor, that is to say, a high proportion of seats occupied, is obviously important, yet I do not want my passengers to feel too crowded."

"It sounds to me as if you are pursuing two mutually incompatible goals there," I said. "You cannot have a high load factor and spare elbow room both together. Your bus manager is being asked to do the impossible."

"I think you are not quite right, Watson," Holmes said, leafing rapidly through the correspondence. "I see that the manager claims that on average 50 percent of the London bus seats are occupied. Is that what you noticed on your own trips, Mr. Rolleman?"

Our guest frowned. "Actually, no. The buses I caught were almost invariably more crowded. I suppose that means my employees have been underdeclaring the passenger load, and skimming off fares for themselves."

"Not necessarily. You see, there is a difference between the average number of passengers on each bus and the average number of fellow passengers (including himself) that each passenger sees. Because on a crowded bus, there are more people present to notice the crowd!"

Rolleman looked puzzled. "I am not quite with you."

"Suppose ten buses, with ten seats each, are carrying fifty passengers altogether. The load factor of the fleet is 50 percent. If the passengers are equally distributed, five on board each bus, the crowding factor that each passenger observes is also 50 percent. Each passenger has a double seat to himself. or herself.

"Now, suppose instead that five of the buses are empty, and five fully loaded. The load factor of the fleet is still 50 percent, the profitability the same. But the crowding factor is now 100 percent: every passenger is on a bus with no spare seats at all. The empty buses are benefiting nobody."

"I see. The optimal ratio of crowding to load factor is when the passengers are distributed between buses as evenly as possible. The more uneven it gets, the more uncomfortable the passengers, at no benefit to the operator."

"Quite so. Now, Mr. Rolleman, I will need a little time to consider the other data you have given me. Shall we meet in a week? Capital, I shall look forward to seeing you then."

When I returned from showing Mr. Rolleman out, Holmes was pouring himself a large whisky.

"Can I get you one, Watson?"

"A little more soda in mine, if you please."

My friend chuckled at my disapproving expression.

"It is a moment to celebrate, Watson. I fancy this new line of management consultancy work will help provide for a comfortable retirement."

"You are planning to charge a good fee? Well, Rolleman can certainly afford it."

"Not only that, Watson." He pointed to the neatly painted model buses that lay all over the floor. "I shall be keeping these. I believe that one day they will become quite valuable."

"These penny toys for children? Really, Holmes!"

"Exactly, Watson: they are given to children. Who will break them, damage them, and lose them. Anything that becomes rare becomes valuable. I would advise you to buy some."

"I am afraid it is rather too unconventional an investment for me, Holmes."

"It is your choice. Well, knowledge is more important than possessions. You will have done well out of the past couple of days if you have learned a moral from the common element present in all the cases we have discussed."

I counted them off on my fingers. "Reports of freak card hands. Fish stocks. Waiting for buses and in shop queues. Really, Holmes, I am blessed if I can see anything in common in that lot!"

"But there is, Watson. Each has to do with the collection of statistics and the limitations of the observer. *How* you observe, and *where* you observe, determine *what* you observe. You cannot build a true picture of the world without remembering that."

I sipped thoughtfully at my drink for a while. "So, a bit like the observer problem in quantum physics, eh?" I said at length.

"No, Watson, as it turned out. In quantum physics, the making of an observation influences the system being observed. Rolleman thought that his employees knew when he was watching, and that his presence affected their behavior. But he was quite wrong; he simply had not thought things through properly. That kind of self-aggrandizement—the idea that what you see is so exceptional that it must all have been rigged for your benefit, in some kind of grand conspiracy—is not only the first step on the road to paranoia, it is the last refuge of the unimaginative. Sometimes I think even the quantum physicists would do well to remember that."

11

The Case of the
Perfect Accountant

HOLMES PASSED OVER THE SALVER on which Mrs. Hudson was accustomed to deposit our mail. One envelope bore handwriting that I knew well. It was from Penbury, my partner in general practice. He is a man I have never particularly taken to, but I have stayed with him on account of one outstanding virtue: he has the greatest respect for Holmes, and so on the not infrequent occasions when I need a stand-in on short notice because Holmes has roped me into some investigation, he is invariably obliging. The contents of the envelope were, as I expected, a second sealed envelope, and I sighed. I became aware that Holmes was regarding me quizzically.

"My dear friend, it is marvelous that even as well as I know you, you retain the ability to puzzle me occasionally." He pointed. "I recognize the handwriting on both the outer and inner envelopes as that of your partner, Penbury. Moreover, I recognize the address on the second envelope, which has been through the post because it is both stamped and canceled, as that of the office you share. Now, why would he be sending letters to himself, and then posting them unopened on to you? I confess I am at a loss."

I held up the envelope ruefully. "This is really evidence of my own foolishness, Holmes. Last week I unwisely goaded Penbury into making a bet with me.

"I was rather ill-advisedly lecturing him on the futility of gambling. In reply he bet me twenty pounds that he could predict the winner of last Saturday's Derby race using what he called scientific formulae. Because I was not due to see him before then, he said that he would write a letter enclosing the name of the winner and then would mail it to me so that I would find it when I arrived at work on the morning of the race. He is an ingenious fellow, Holmes, and I was rather relieved to find no such letter on the mat when I went in on Saturday. But the note enclosed here tells me that the letter was delivered late: Penbury himself found it on Monday morning. The envelope I hold is clearly postmarked Friday, so if it does contain the winner's name, I am morally bound to pay off on the bet."

"And do you not trust Penbury?"

I wriggled slightly under Holmes's knowing gaze. "In medical matters, yes, of course, or I would not be his partner. But he has a slightly twisted personal morality, or perhaps I should call it a warped sense of humor. He takes the view that if men who deal with him are foolish, then he is within his rights to take advantage of that foolishness, as long as he himself is not guilty of any direct lie or illegal act."

"Hmm! That sounds like the self-justification of an amateur con man studying to be a professional. You are quite right to be suspicious. You think he is capable of having tampered with the envelope? Pass it across to me."

I did so. Holmes examined the edges of the envelope minutely through a hand magnifier and then scraped delicately at them with a small surgical knife. He used an eyedropper to add some chemical at the edge of the seal and then tested it again. He shook his head in disappointment and passed it back to me.

"It is intact as far as I can tell, Watson. Standard stock, waterproof glue, no sign of its being opened and resealed, and the postmark is authentic."

I opened the envelope and sighed. "Michelangelo, the winner, sure enough. That is twenty pounds of my hard-earned cash gone. Oh well, at least it will teach me to be less boastful in future."

Holmes nodded and continued to open his own correspondence. Suddenly he smote his forehead. "Good heavens, I am slow today, Watson. Have no fear, your twenty pounds is safe."

I looked at him in puzzlement and then examined the envelope and letter again.

"I thought you said the envelope had not been tampered with."

"Nor has it, Watson. The answer lies not in the letter that you see, but in those that you do not.

"There were eight runners in the race. All Penbury had to do was to post eight letters on Friday, each with the name of a different horse. He posted them sufficiently late to ensure that they would be stamped with that day's postmark, yet not be in time for Saturday delivery. On Monday he called in at the surgery and found all eight letters fresh on the mat as he expected. He could tell which envelope was which because he had marked them in some way, probably by spacing out the address slightly differently on each so that no individual envelope looks in any way suspicious. Then he burned the envelopes of the seven losers and forwarded the remaining one to you."

"Confound it, Holmes! But if he has burned the other envelopes, how will I prove it?"

"You describe his personal morality as preventing him from lying outright—or at least, he wishes you to think he will not stoop to that. If you ask him directly, I think he will admit it."

"I am most grateful, Holmes. I must apologize for asking you to exert your powers on so trivial a matter."

"Not at all, Watson. In fact, I wish you had consulted me before."

"Why so?"

Holmes pointed ruefully at the small pile of letters and the large heap of account books that Rolleman had left him, which he had been poring over for most of the past week. "It would have given me a valuable hint. I have wasted a lot of time checking the claims made in these letters, all from junior managers to Rolleman boasting of specific achievements, against the full accounting records. The claims are all true; of course the junior managers would not dare to be caught in a direct lie, and I should have seen that. The significance is rather in the claims they do *not* make. It reminds me a little of that case of the dog that did nothing in the night."

"Well, it sounds as though you have made good progress."

Holmes shook his head angrily. "In that department, yes. But in the more serious matter, I have been banging my head against a brick wall."

He pointed at the green-bound sales ledgers I recognized as containing the Barnum's accounts.

"I am sure that Rolleman's suspicions about his great department store are correct. Turnover has been falling rapidly, judging by these figures. Yet one need only visit the shop to see that business is booming." He drummed his fingers. "But I can find no errors or inconsistencies in these accounts, Watson! Purchases, sales, stock: all the figures cross-check perfectly."

"What can the solution be, then?"

"Why, only that these accounts are pure fiction! They have been invented from scratch and tell a self-consistent story that has no relation to the reality of the Barnum's transactions."

I looked at him severely. "Obviously some legwork is called for, Holmes. It is no good sitting here in your dressing gown;

you must get out and investigate in person. It will do your constitution good. In truth I have been a little concerned to see you stewing indoors so much with the weather so fine. If those figures are invented, you cannot possibly prove it just by looking at them."

"You are wrong there, Watson. Remember the case of Madam Zelda? It is very difficult to invent figures without including some telltale pattern that gives the game away."

He picked up one of the ledgers, reviewed it briefly, and then put it down again with a sigh. "But whoever composed these is a great deal cleverer than Madam Zelda's cousin. So far I have drawn a blank. I shall take your advice, Watson. I shall take a bath and then venture forth to see what I can find out with my eyes and ears."

Knowing Holmes, I was not taken aback when half an hour later his door opened to reveal a ginger-haired man with gold-rimmed spectacles and a neatly trimmed moustache, immediately identifiable as a manufacturer's representative. I regarded him critically.

"The hair perhaps a little more neatly combed, Holmes?"

"I think not, Watson. I have supposedly been walking the streets of London all morning, visiting different shops to drum up trade." His gaze fell to the ledger and notepad on my lap. "I see you have been doing some analysis of your own."

I gave a self-deprecating laugh. "Yes, I thought for a moment I might be on to something. But I was completely mistaken: it is really not worth going into."

"Come, Watson: you know I am ever eager to hear your thoughts!"

I was forced reluctantly to continue. "I thought I would go down the figures for the individual sales and look at the first digit of each number. You once told me that a person who is asked to pick a digit between 1 and 10 very often picks 7. I wanted to see whether the first digits were indeed random or showed some kind of bias.

"I started by counting the 1's, and found the incidence was not 1 in 10, as I expected, but about 1 in 9. The same was true of the 2's and the 3's. I thought it quite suspicious. Then I realized that a price never starts with 0, so actually, 1 in 9 was the frequency to be expected for each digit," I confessed, red-faced.

Holmes gave a snort of laughter. Then he hesitated. He seemed to be staring into space, and his lips moved silently. Then he smote himself on the forehead. He peeled off the false moustache and flung it on the side table.

"Congratulations, Watson! You have saved me an onerous day's work. I knew in my heart that there was something deeply suspicious about these figures, and you have put your finger on it."

"But I have just admitted I see nothing suspicious."

"Doubtless not, Watson, but I congratulate you anyway. Now I can devote my day to one or two violin passages I have been neglecting to perfect."

"You mean you are not going out after all?" I asked in dismay.

Holmes smiled. "You are as ever concerned about my health. We shall compromise. Rolleman expects to meet me today: we can have lunch in the open air, by the bandstand in the park, and we shall walk there together. That should satisfy you that my constitution is being maintained."

To my relief, the bandstand was not actually in use, and the fresh breeze of the park was really very refreshing after the heat of the London streets. We took our seats at a round table with a flapping tablecloth that kept threatening to blow away. Sherlock Holmes placed the ledgers he had brought strategically to weight it down—quite the best use for a dry book of figures, I thought!—and also set an artist's sketchpad in the center of the table.

"The good news I have for you," he said as he handed the bundle of correspondence back to Rolleman, "is that all the claims made in these letters are true. The bad news is that they are all misleading. For example, examine the top three."

Rolleman did so. "These managers are all claiming the same thing: that the average wages in their departments have increased in the past year," he said. "The first by an impressive 18 percent, the second and third by a tremendous 40 percent. And they have done this without affecting the profitability of their departments! My goal is that salaries should be fairer, the income more evenly distributed. I consider in particular that the wages my father paid to his lowest-ranked employees were a disgrace. Clearly, these managers have worked to remedy the situation, and the first should be praised, the others praised and promoted."

"No, sir, the three managers are not claiming the same thing at all. The first correspondent says that the *mean* wage has increased. The second says that the *modal* wage has increased. The third says that the *median* wage has increased."

Rolleman frowned. "I seem to recall a mathematics teacher telling me that those often mean the same thing," he said.

"Not at all: they are defined in completely different ways. The mean wage is simply the total wage bill divided by the number of employees. The modal wage is that salary band in which the largest number of employees falls. And the median wage is defined as follows: line the employees up in order of the size of their salaries. The median is the salary of the man or woman who comes exactly halfway down the line.

"Now in the case of the famous Normal Distribution—the so-called bell curve, which you aptly dubbed the Napoleon's Hat curve, Watson—the three values are in fact the same. For example, if you made Mr. Rolleman's employees queue up in order of height, you would observe that the mean height turns out also to be the most common height, which is also the height of the man halfway down the queue.

"But a graph of salaries tells a completely different story. Here is a bar chart showing how your father allocated wages

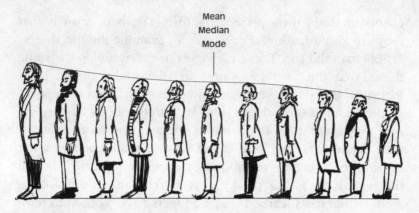

Mean
Median
Mode

Employees by Height

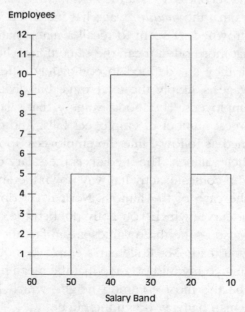

Employees by Salary

in a typical department. There would be one manager on a salary of $55, five supervisors on $44 each, ten skilled workers on $35, twelve semiskilled on $25, and five apprentices earning $15.

"The total wages bill is $1,000. The number of employees is 33. The mean salary is the first divided by the second: near enough $30. The modal salary, on the other hand, is $25, and the median—the salary of the seventeenth employee—is also $25.

"Now, what do you know has occurred, if the mean salary has increased?"

Rolleman scowled. "Either the total wages bill has gone up, or the number of employees has gone down," he said.

"And what does that tell you about the fairness of the share that each receives?"

"Why, nothing at all!"

"Quite so. For example, if the manager sacks all the apprentices and adds the $75 saved to his own salary, then average earnings have gone up by 18 percent. That is precisely what the manager of your Iowa office has done."

"Why, I will have his hide!"

"I should hope so. Now, starting from the original picture, let us arbitrarily demote three of the supervisors to skilled status, and add the $27 saved to the manager's salary to keep the total wage bill constant. The mean salary is therefore unchanged. But the modal salary increases 40 percent, from $25 to $35! That is just what your Delaware manager has done."

"I shall have his hide also!"

"Now to the New Orleans office. The median is usually a somewhat more reliable indicator of fairness, but it too can be abused. In the original picture, promote one semiskilled worker to skilled, and save the $10 by docking $2 from each apprentice. The mean is unchanged, but the median increases from $25 to $35, again by 40 percent."

I could contain myself no longer. "I have heard it said that there are 'lies, damn lies, and statistics,'" I said, "and now I see that it is true. Your countrymen have a phrase, Mr. Rolleman: 'I'm from Missouri, I'll see for myself.' Clearly, one must take no account of statistics but, rather, travel and observe in person."

Sherlock Holmes shook his head gently. "Not so, Watson. Statistics can be deceptive, but they are, to use another Americanism, 'the only game in town.' We saw last week how personal observations can also be misleading. You cannot possibly gain a fair picture of what is going on in a big business, still less a whole nation, from one viewpoint only.

"Statistics are essential. The point is that you still can be deceived if you rely on selected data—that is, data selected by somebody else. You need to see the whole picture for yourself. That is best done by looking at a graph or chart that embodies all the data you are interested in. The human visual system is exquisitely good at detecting patterns, Mr. Rolleman: if your managers had sent you bar charts of their new salary structures, you would have seen instantly what they were up to.

"By the way, a graph is also vital if you want to see clearly how something is evolving with time. I will pick two examples." He held up another letter. "The manager of a factory in New Jersey that makes matches has pointed out that although sales are still declining, they are doing so at a slower rate than last year."

"That does sound like an achievement of sorts, if a modest one," I ventured.

Holmes showed us the picture his correspondent had included. "Actually, you can see here that sales are declining exponentially—that is to say, by a fixed percentage every month. That is about as bad as things could get, but because the slope of the graph, which indicates the rate of decline, becomes ever shallower, the manager can always tell you that the decline is showing. The slope of a graph is always an interesting property. Let me teach you calculus—"

Here I leapt up in alarm, overturning my chair. "I have a horror of algebra!" I cried.

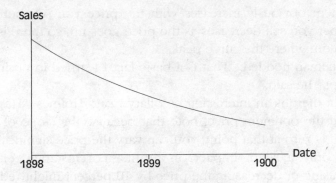

An Exponential Decline

"Sit down, Watson. I will do it visually. Nearly all of the practical utility of the calculus derives from the simple observation that at the maximum or minimum point of a curve, its slope must be zero." He drew another graph.

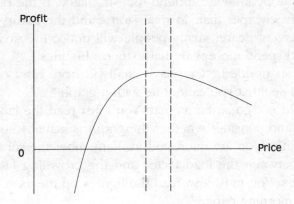

Price Versus Profit

"That is trivially obvious," I said.

"So, find the point of zero slope, and in many contexts you find the optimum. For instance, this curve could represent the price of a product versus the total profit on sales. The profit per item obviously increases with the price you set, but the number you sell decreases as the price goes up, so there is an optimum where the curve peaks."

Rolleman nodded. "That is a basic fact I learned in business school," he said.

"But there is an interesting corollary," said Holmes. "Having found the optimum price, note that because the slope of the graph is zero at that point, you can vary the price significantly with only a slight impact on profits. For instance, either increasing or decreasing the price by 10 percent might reduce profit by only 1 percent." He sketched the dotted lines.

"If your social conscience troubles you, Mr. Rolleman, vary your supermarket prices as follows: make the bread a little cheaper and the luxury items a little more expensive. The poor will benefit from the difference, and at almost no cost to yourself.

"You might even benefit the store, indirectly. You are familiar with the penny-wise, pound-foolish fallacy. If the bread is 20 percent cheaper than in rival stores and the luxury items only 5 percent dearer, many people will notice the saving on bread and spend more than that extra on luxuries."

Rolleman nodded. "Graphs it shall be from now on, Mr. Holmes. I see that you cannot lie with a graph."

"I would not go so far as that. You must read the labels on the axes and consider whether the graph is actually showing the quantities you are interested in. Remember the subtle difference between the load factor and the crowding factor of your buses. You must also read the figures on the axes. May I see your morning paper?"

Rolleman looked surprised at the sudden request but passed it over. Holmes leafed through the pages.

"Here is a perfect example of how journalists fail to describe numbers with words. The writer tells us that there was an appallingly high incidence of parasitic infections in Gambezian children: 40 percent had some form of illness. But in the last year there has been 'a 10 percent improvement.' Does he mean that the fraction of healthy children has increased a tenth, from 60 to 66 percent? Or that the fraction of infected children has fallen a tenth, from 40 to 36 percent? Or that the fraction of infected children has fallen from 40 to 30 percent? There is no way to tell, and I very much doubt that the writer himself either knows or cares about the difference. Any reasonable graph or pie chart would convey the information unambiguously, but of course there is none."

He turned the page. "Ha! Here is a perfect example of misleading graphs." He held a page out for us to see. "Deaths by leopard and crocodile versus deaths from disease in Gambezia."

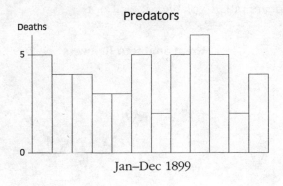

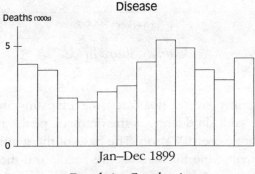

Death in Gambezia

"It must be strange to live in a country where you are as likely to die in the jaws of a predator as from an illness," I remarked.

"Nonsense, Watson. Look at the scales at the side! Comparing them reveals that a thousand times as many people die from disease as from predators, even in central Africa. But the writer knows that big cats and crocodiles are more exciting, so the graph of that relatively trivial risk has been blown up to the same scale as the larger one. Newspapers always report lurid deaths more prominently than mundane ones, so our perception of risk gets very biased. In reality, the fly and the mosquito kill far more people than the lion and the tiger.

"Here is a further example of a misleading graph, this one from the financial pages." He turned the page. "This stock price appears rather volatile, does it not?"

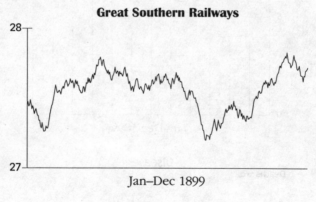

Great Southern Railways

Jan–Dec 1899

Railway Share Price

"You certainly would not catch me investing in Southern Railways," I said, shuddering at the dramatic peaks and valleys.

"Look more closely, Watson. The base of the graph is not set at zero. Actually, the fluctuations are small, and the stock has stayed very near 27 1/2 pence a share throughout the year."

"So the moral is: be wary, but a graph is still infinitely better than words, when describing statistics," said Rolleman thoughtfully. "I shall make sure that all my departments use that method of reporting in future."

A waiter approached us with menus. Rolleman waved him away.

"This is all very interesting, Mr. Holmes, but before we eat there is still one thing on my mind. Were you able to find the problem with the Barnum's accounts? My best accountant—a fellow I brought over from New York—keeps telling me something looks strange about them, but he is darned if he can put his finger on it."

Holmes waved an airy hand. "Oh, yes. That was very obvious. Why, Watson here spotted it at once."

Rolleman gazed at me with an incredulity that I found rather insulting.

"He noticed that the digits from 1 to 9 appear about equally often as the first digit of each entry."

"But surely that is to be expected?"

"Not at all. The digit 1 should be far more common. I can tell you that almost a third of the price labels in your store start with the digit one."

Rolleman frowned. "Perhaps you got that impression wandering around the shop, but I think your intuition has deceived you. The store has thousands of lines, and the law of averages must surely apply. You are talking nonsense, Mr. Holmes."

Instead of becoming annoyed, my friend smiled impishly. He glanced down at the menu in front of him.

"The prices in London really are becoming absurd. But if you think the food is dear, look at the wine list. The champagne especially; even for Bollinger's, that is an outrageous sum."

He met Rolleman's eye. "Mr. Rolleman, I will bet you a bottle of this champagne that I can demonstrate that over the coming century, the price label on any product you name in your store will start with the digit 1 nearly one-third of the time."

"Are you feeling quite well, Mr. Holmes?"

"I mean it: pick any item of your choice."

"Very well: a bottle of our best perfume. That is 30 cents in New York; I am not sure of the exact price here."

"We will work in dollars; it will make the calculations easier in any case. Watson, have you your pocket calculator with you?"

I produced my slide rule. Every Victorian schoolchild must be familiar with the device, but in case any of my readers live in primitive countries where it is unavailable, I have sketched it here.

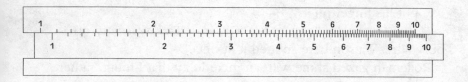

The Slide Rule

"Mr. Rolleman, you are familiar with the process of inflation: the dollar is worth a little less each year. What would you guess the inflation rate over the coming century will be? The precise value does not matter: simply choose a reasonable figure."

"Let us say 7 percent per annum."

"Very good. Watson, be so good as to make a pencil mark at 3 and multiply by 1.07."

I did so. My sketch shows the rule in the appropriate position to multiply any number by 1.07. For example, reading from the bottom scale to the top, you can see that 1.07 multiplied by 3 is 3.21.

"It is 32 cents to the nearest whole cent," I said.

"That will be its price next year. Be so good as to make a second pencil mark and multiply by 1.07 again."

I did so. "In 1902, it will cost 34 cents."

"And again, if you please."

I did so. "In 1903, 37 cents. The pencil marks are spaced an even distance apart," I remarked as we continued.

"Of course, Watson. A slide rule is graduated in what is called a logarithmic scale, as opposed to the linear scale you see on a standard ruler. Traversing a certain distance always multiplies by a corresponding factor. That is why a slide rule works as it does. Going one complete circuit multiplies by a factor of 10; of course, you must remember to wrap from 10 back to 1, and to move the decimal point, when you reach the end of the rule."

As I performed the calculations with the slide rule, I listed the results. In 1918, the perfume would cost $1. In 1934, $2.99 ... "Why, we have come back almost to our starting point," I said.

The Increasing Price of Perfume

1900 $ 0.30	1919 $ 1.08		1962 $ 19.90	1982 $ 77.02
1901 $ 0.32	1920 $ 1.16	1942 $ 5.14		
1902 $ 0.34	1921 $ 1.24	1943 $ 5.50	1963 $ 21.30	1983 $ 82.41
1903 $ 0.37	1922 $ 1.33	1944 $ 5.89	1964 $ 22.79	1984 $ 88.18
1904 $ 0.39	1923 $ 1.42		1965 $ 24.38	
	1924 $ 1.52	1945 $ 6.30	1966 $ 26.09	1985 $ 94.35
1905 $ 0.42	1925 $ 1.63	1946 $ 6.74	1967 $ 27.91	
1906 $ 0.45	1926 $ 1.74		1968 $ 29.87	1986 $ 100.95
1907 $ 0.48	1927 $ 1.86	1947 $ 7.21		1987 $ 108.02
	1928 $ 1.99	1948 $ 7.72	1969 $ 31.96	1988 $ 115.58
1908 $ 0.52			1970 $ 34.20	1989 $ 123.67
1909 $ 0.55	1929 $ 2.13	1949 $ 8.26	1971 $ 36.59	1990 $ 132.33
1910 $ 0.59	1930 $ 2.28	1950 $ 8.84	1972 $ 39.15	1991 $ 141.59
	1931 $ 2.44			1992 $ 151.51
1911 $ 0.63	1932 $ 2.61	1951 $ .9.46	1973 $ 41.89	1993 $ 162.11
1912 $ 0.68	1933 $ 2.80		1974 $ 44.83	1994 $ 173.46
	1934 $ 2.99	1952 $ 10.12	1975 $ 47.96	1995 $ 185.60
1913 $ 0.72		1953 $ 10.83		1996 $ 198.59
1914 $ 0.77	1935 $ 3.20	1954 $ 11.58	1976 $ 51.32	
	1936 $ 3.43	1955 $ 12.39	1977 $ 54.91	1997 $ 212.49
1915 $ 0.83	1937 $ 3.67	1956 $ 13.26	1978 $ 58.76	1998 $ 227.37
1916 $ 0.89	1938 $ 3.92	1957 $ 14.19		1999 $ 243.28
		1958 $ 15.18	1979 $ 62.87	2000 $ 260.31
1917 $ 0.95	1939 $ 4.20	1959 $ 16.25	1980 $ 67.27	
	1940 $ 4.49	1960 $ 17.38		
1918 $ 1.01	1941 $ 4.81	1961 $ 18.60	1981 $ 71.98	

"The perfume is now roughly ten times as expensive as before. You can stop there, Watson—oh, very well, continue if you must."

"In the year 2000, this perfume will cost $260 a bottle!" I said eventually in horror.

"It will not seem quite so much by then, Watson. Now, start at the 1900 value, and tell me how often each successive initial digit is hit."

"The 3 comes up five times. The 4 and 5 three times. The 6, 7, and 8 twice each. The 9 only once."

"And then the 1?"

"Eleven times!" I exclaimed.

"Exactly so, Watson. The point is that starting at $1, the digit does not change until the perfume has doubled in price. But starting at $9, the digit changes when the price has increased a mere 12 percent. The same happens at $90, $100, and so on. You always wind quickly past the high digits and then get stuck in the low ones. In fact, you can see that because the pencil marks are evenly spaced along your slide rule, the probability of each digit occurring is just in proportion to the length it occupies on the rule. About 30 percent for 1, down to 5 percent for 9."

Rolleman considered. "I can see that if you go into my store to purchase a particular product in a randomly chosen year, the price is likely to start with a low digit. Yet I am still not quite convinced that the pattern will hold for a comparison of all the current stock at a particular time."

"Look at it this way, then. The low-digit ratio holds for the list I have made because the price increases not by a fixed amount each year, but in geometric ratio—that is, by an amount proportional to the current price."

"Like compound interest; quite so."

"Imagine you took every item in your shop today and arranged them all in consecutive order of price, from the cheapest to the most expensive. You would find that as you went along the line, the prices similarly grew in geometric proportion: the difference in price becomes larger as the price becomes larger.

"For example, I spent some time in your bathroom-fittings section, and I noticed there is a choice of taps: brass at 2 cents, bronze at 3 cents, silver at 4 cents."

"That is typical of my stocking policy. I like to offer a choice, with the most expensive about twice as dear at the least."

"You also stock bathroom suites, starting at $10, with the most expensive at $21. Now tell me, do you stock a bathroom suite costing $10.01, $10.02, and so on up to the dearest? Of course you do not. It would be absurd, and there would not be room in the store for the stock. Prices increase by a ratio, not by a fixed amount.

"The law of the most common first digit was discovered by one Simon Newcomb in 1881, although to my mind he never quite explained the reason for it, and it remains rather obscure. But it is a very general law, applying not just to man-made objects. Collect almost any population that varies greatly in size"—he waved an arm—"say, the boulders at the foot of a glacier. Arrange them in line in order of size. You will find that the difference in size is greater for the larger boulders than for the smaller. Otherwise, there would be unreasonably many large boulders and unreasonably few small ones. The Universe simply has more room for small objects than for large ones.

"Newcomb's law is a complement to the famous Normal Distribution, or bell curve. If you take any population that is relatively uniform in magnitude, such as humans measured by height, often it is a good fit to the bell curve. But if you take a population of objects that vary from the smallest to the largest by many orders of magnitude, that is, many factors of 10, it tends to conform to Newcomb's law.

"One could discuss the philosophical implications further, but I am a practical man, and happy to stop at having demonstrated the guilt of your shop manager. My throat has become quite dry," Holmes said meaningfully.

Rolleman grinned, and snapped his fingers at the waiter. "I am quite happy to buy the champagne, gentlemen. We will toast a powerful new method of discovering fraudulent accounts!"

"What have you there, Watson? Not more correspondence from Penbury, I hope."

I gave a guilty start. "Oh, no. As you predicted, he owned up when challenged," I said. "In fact, he was very good-humored about the whole thing. He said it had been a little test of your powers, and he assured me he would have come clean in any event. Clearly I misjudged him."

Holmes snorted. "That is far from clear," he said. "We have no way of knowing whether he would really have refrained from collecting if his subterfuge had not been spotted. But you have not answered my question, Watson."

"Well, it is a package I must confess I find quite exciting. A week ago I received a letter with a certificate attesting to the fact that my name had been drawn from a hat to go forward into the second round of a competition to receive a worth-while set of prizes." I looked down at the letter before me. "I do not quite know how my name came to be entered, but the rewards range from a guaranteed tax-free income for life, through a free trip to America by luxury liner, down to conso-lation prizes such as stationery.

"Now I am told that my name has gone through to the final round! I have therefore won a guaranteed prize. All I have to do is take out a year's subscription to the magazine sponsoring the competition and await the result of the draw."

I looked at my colleague's disapproving face. "Now, Holmes, even you admit it is sometimes good sense to gam-ble. I do not believe the letter is a lie. The subscription is to a famous magazine, which would be in legal trouble if any kind of fraud were involved."

I held up an impressive second certificate, confirming that I was guaranteed to be a prizewinner and bearing the printed signature of the editor of the magazine involved. "Having been so lucky already, surely I can assume the odds are in my favor."

Holmes shook his head. "Watson, Watson, when will you learn? You do not have to lie to deceive people about statistics. You have merely to tell selected truths. The letter does not tell you how many names were selected out at each stage. It does not tell you how many stationery prizes, relative to lifetime incomes and holidays, are to be given out.

"My guess would be that 100,000 names went into the first round and that 99,900 were selected as 'winners' for the second. A similar proportion were informed that they had become finalists. If you take out a five-pound lifetime subscription to the magazine, you will stand perhaps 1 chance in 100,000 of winning a lifetime income, 2 chances in 100,000 of enjoying a free holiday, and a virtual certainty of receiving a cheap pen prominently embossed with the name of the sponsor. The letter is strictly true—the senders would no doubt have me up in court if I asserted otherwise—but at the same time it is highly deceptive. If Barnum Rolleman were here, he would tell you the Americans have a name for this kind of thing. They call it *junk mail*."

I sighed. Somehow I had known it was too good to be true, but I had been on the verge of signing my check to the magazine. I crumpled up the whole bundle of paper in disgust. "I suppose there is no use for this at all, then?" I said bitterly.

"I did not say that. Toss it in the fireplace. Autumn is not so far away, and if the phenomenon continues, at least we shall not want for fire-lighting material."

12

Three Cases of
Good Intentions

THE DEAD MAN SEEMED TO stare at me in a most disconcerting way. Eyes open in a wide-mouthed expression that combined horror, surprise, and fury, the late Eli Squires did not appear to have gone willingly to meet his maker. Staring into that evil face, I could quite understand Sherlock Holmes's skepticism that the recent wave of suicides among London's criminal classes were truly motivated by remorse!

I watched my companion move swiftly about the shabby little room in which Eli had lived and died. He paid close attention to half a dozen butterscotch sweets that lay scattered on the table before the corpse. I could see nothing significant about them at all. Then he picked up a sheet of paper beside them.

"'I, Eli Squires, admit to the forgery of five hundred pounds in fake Treasury notes and have decided to pay the price,'" he read aloud.

"That is all he has felt the need to write, apart from his signature and the date. All written in a firm hand matching that on his various records and accounts. No signs of violence upon the corpse. It is really most singular, Watson."

"Perhaps he felt the police net closing in upon him," I suggested.

Holmes nodded. "As indeed it was. But Eli was no quitter. He had money, contacts, and the ability to forge any kind of document he needed. Why did he not run? He could have booked passage on a steamship to any foreign destination under any name he chose. Or merely fled to another part of this country and altered his identity. It is inconceivable that he should simply decide to give up, at this stage of the game."

"You think he was murdered, then. But how?"

By way of answer, Sherlock Holmes moved behind Eli's chair and leaned forward to wrap his arms about the body's waist. Without warning, he tightened his grip with considerable force. The upper part of the corpse moved, the face changing, the mouth gaping still wider. An awful groan issued forth, and from its lips shot a small, round object that almost struck me in the face. I drew back with a cry of disgust.

Holmes sprang forward and picked the object up in his handkerchief. He held it down beside the sweets on the table; it was recognizably the same, but with its outer layer dissolved away.

"He was poisoned, then?"

"I believe so."

"Then the mystery is explained. Someone gave him an apparently innocent gift that turned out to be deadly."

"Really? Then how do you account for the note, Watson?"

I was left rather at a loss.

"I suppose he must have been forced to eat the sweet," I said.

Holmes hesitated. "So it would seem," he said at last. "And yet there is no sign of a struggle. Eli must have known what was coming. Even at gunpoint, why should he have given in so easily? No, Watson, there is some deep mystery here."

At this point heavy footsteps sounded on the stairs outside the door. Holmes hastily thrust the handkerchief and its contents into his pocket. A moment later Lestrade and two uniformed constables had joined us.

"Gentlemen, I see this time you are ahead of the Force," said Lestrade good-naturedly. "Well, we beat you to the other three cases, so you are quite welcome to your turn. Really, if the criminals of London continue topping themselves at this rate, I will be able to put my feet up. No one on the Force will be mourning this gentleman, any more than they did the other three."

"All four were known to you, then?" I asked.

"Oh, dear me, yes, Doctor. Not only were they known to us, but they had all been charged with crimes. They were only out on bail while our new Director of Prosecutions made up his mind whether to proceed with their cases. Albert Forsyth, burglary, should have got two years. Derek Carstairs, robbery with violence, expected to get ten. Julia Neilson, running a vice ring, likely to go down for four. And now Eli Squires, small-time forger; I had put him down for five. That is, if our bewigged Director had kept his nerve. To my mind he has been refusing to prosecute cases where the evidence was quite clear, and a fair number of criminals have been going scot-free. Delegating police powers to lawyers was bound to be a disaster; I said so at the time."

"To what do you attribute their deaths?" I asked.

Lestrade hesitated. "Well, I am split in my mind between suicide and murder at the moment. It could be suicide. But my betting is that it is part of some turf war between criminal gangs, fought with rules we do not quite understand. Frankly, London is well rid of them all, and I for one will shed no tears if the matter is never cleared up. I shall do my duty and investigate, of course. But frankly, Mr. Holmes, I do not see why you are bothering yourself."

Sherlock Holmes did not reply directly, but pointed to the table. "Note the butterscotch sweets, again," he said.

Lestrade tossed back his head and roared with laughter. "Your colleague has a positive obsession with confectionery," he told me cheerfully. "Yes, Mr. Holmes, we saw similar sweets at the scenes of all the other deaths. And yes, I have had them all analyzed. And no, none of them contained anything other than sugar and such harmless ingredients. If you do not trust our laboratory, Mr. Holmes, by all means help yourself to one or two."

Holmes took two of the sweets with thanks. I noted, however, that he made no mention of the more gruesome remnant wrapped in his handkerchief. Soon we were striding back toward Baker Street.

"What was all that about this Director?" I asked.

Holmes smiled. "Lestrade is engaged in fighting a small turf war of his own," he said. "In Scotland, the decision whether to prosecute a given criminal is made by a Crown servant known as the Procurator Fiscal, who is independent of the police. It is his job to assess the probability of a conviction being obtained, and to proceed only if he is satisfied that the chance of success is reasonably high. Here in England, the police have always made their own decisions on prosecution. But there is an experiment running in the London area this year to see whether appointing a similar independent referee—in fact, the new Director of Prosecutions is the well-known barrister the Honorable George White, QC—can save public money and achieve a better success rate in the prosecutions that do go ahead. Lestrade is jealous of his powers being eroded."

He shook his head. "However, I must say that it does seem to me also that White has been prosecuting far fewer cases than he could. Perhaps he is overmindful of saving public money, although that would be a strange trait in a lawyer."

"I seem to remember he has campaigned for prison reform," I said. "Perhaps his conscience will not let him proceed with a case unless he is certain no injustice will be done."

"Possibly. Although if the authorities are too lenient, it can be self-defeating: sometimes vigilantes take matters into their own hands, bypassing even the imperfect safeguards by which our legal system tries to prevent the innocent being wrongly punished."

"Do you think that has happened in these cases?"

"It seems possible. It is puzzling at one level, because the alleged crimes are not of the usual sort to excite such attacks. But the newspapers did publish details of the charges against each of those who have died so strangely."

We were entering our rooms as he spoke, and he picked up his most recent scrapbook of newspaper clippings. "Here is an example from *The Times:* 'Albert Forsyth, 51, of Cottisham Road, Hackney, was yesterday charged with burglary.' That was all; they are not permitted to say anything that might prejudice the trial. Then, two weeks later, 'Albert Forsyth, 51, was today found dead at his lodgings. Nine butterscotch sweets of unknown manufacture were found on the table before him; food poisoning is suspected.'

"Curious that newspaper reports always tend to give the person's age," I said. "I suppose that if the person involved is not famous, the point they have reached in their passage through life is at least something fundamental to help readers visualize them."

"The other reports are very similar. You are right, Watson, they all give the age. 'Derek Carstairs, 29.' He was found with three sweets in front of him; they mention the food poisoning idea more prominently. 'Julia Neilson, 45.' She left five sweets, but there is mention that the sweets have all been tested and none found to be harmful. 'Eli Squires, 34, of Hatton Mews, charged with forgery.' There is no corresponding death notice yet, of course; it will be in tomorrow's papers."

Holmes rolled up his sleeves. "Now I must get to work, Watson. I am determined to identify the poison in the sweets that those bunglers at Scotland Yard's laboratory have missed."

I turned my attention to a set of brightly colored booklets that I had been meaning to read for some time. Those first months of the twentieth century had been marked by a positive explosion of ideas for a new and better society. Remarkable visions of a world without war, without crime, without poverty or oppression, were being published. And not by cranks—such brilliant and famous men as George Bernard Shaw and H. G. Wells were among those authoring and endorsing the pamphlets in which these bold forecasts appeared. I felt it was my duty to try to understand them before this brave new world was upon us.

Soon fantastic images were dancing in my head. In the past century I had seen the horrors of war at first hand. I daily witnessed the suffering of extreme poverty, even here in the capital of the most powerful Empire the world had ever seen. How wonderful it would be when these things were no more!

I dimly understood that just as Sherlock Holmes had lectured me on how my life would be improved if I learned to make my decisions more scientifically (how long ago that seemed—and no wonder, for in a manner of speaking it had been a century ago!), so these authors wanted to apply the principles of logical and scientific decision making to government, to economics, to the machinery of justice. And the result would be Utopia. Alas, some of the details of just how this would be achieved seemed to be beyond me, and as I struggled with the concepts, I am afraid I must eventually have fallen asleep.

I awoke to see Sherlock Holmes answering the door to an errand boy. He slit open the envelope offered. "Well, this is a surprise, Watson. Brother Mycroft has invited us both to tea."

"It is good see family ties are being maintained and strengthened, Holmes."

He smiled. "There is a reason for everything Mycroft does, Watson. He encloses a peculiar request: he wishes me to bring three of the more presentable of the Baker Street irregulars. We will doubtless discover more when we get there. Still, there is something in what you say. I will accept, on the condition that Mycroft join us for supper here afterward. Remind me to tell Mrs. Hudson to expect one extra at dinner."

He scribbled a note for the errand boy, who left at a run. I noticed that my friend's hands were heavily stained with chemicals.

"How went the analysis of the butterscotch?"

My friend smiled ruefully. "Not quite as I expected. First I tested the sweet from Squires's mouth: it contained a massive dose of cyanide. It seemed incredible that the Scotland Yard chemists could have missed something so obvious. But then I tested the two intact sweets. There was no trace of cyanide. Nor have I been able to find any sign of any other untoward substance."

"How extraordinary," I agreed. But a moment later I snapped my fingers. "I have it, Holmes. A sadistic murderer, just as one would expect of some evil underworld hit man. He arrives with a bag of a dozen sweets. He tells the victim one is poisoned, and forces him to down them one by one until the poisoned sweet is reached in a diabolical form of Russian Roulette. The killer leaves the uneaten sweets behind, and of course by the laws of chance, the number varies."

"You could be right, Watson. In fact I will go so far as to say you must be partially right. But there is still something here that eludes me. Let me think about it as we walk."

"You are quiet, today, Watson," Holmes observed as we turned into the Mall.

I shrugged my shoulders. "I am rather pensive, but it is nothing to trouble you with. Several of my patients have died unexpectedly in the past few weeks."

"In suspicious circumstances?"

"Good lord, no! They were all in the last stages of Baird's disease. The deaths were quite predictable."

Holmes looked at me searchingly. "There is still something on your mind, Watson."

"Well, I admit it was disappointing. You see I recently discovered, quite by chance, that an extract of nightshade long used to treat other conditions seemed to be an effective cure for Baird's. Penbury and I have been quite excited: if true, it is an important discovery and will be worth a paper in *The Lancet* at least. But the treatment has been less effective than the early results suggested. I have now had dozens of Baird's cases referred to me. I prescribe nightshade extract every time, and in half the cases the recovery is little short of miraculous. But in the remaining cases there really seems to be no effect at all: the patients die as rapidly as if they had gone untreated."

Holmes looked thoughtful. "To which chemists do they take your prescriptions?"

"I have thought of that, Holmes: some pharmacists are more reliable than others. But in the present instance I can be sure that all the patients are receiving exactly the same dose, for Penbury acts as pharmacist and makes up the medicine for me. I hand every patient an identical bottle of the little blue pills, but they work in only half the cases."

My friend shrugged. "You know better than I, Watson, that patients can react to the same medication in surprisingly different ways. You have my sympathy, but we must all harden ourselves to occasional failures."

We must have been an odd sight walking up Whitehall: Holmes and myself in our best clothes, but followed at a short distance by three of the ragged Baker Street urchins, as Mycroft had requested. Immediately after entering the intimidating Foreign Office buildings, the lads were shown to a side corridor, and Holmes and I were ushered into a magnificently tall-ceilinged and chandeliered room that would not have disgraced one of London's best clubs. Upon the wall-to-wall deep-pile carpet, four armchairs faced one another in the center; to the side was a tea table on which reposed several large cakes. Mycroft joined us almost immediately.

"It was good of you to come, Sherlock, Doctor," he boomed. "As we are to socialize later, let us get down to business straight away. I had several reasons for inviting you, but first of all I need your advice regarding the Balkans—especially yours, Doctor."

I thought that London's greatest mind must have come unhinged. If his own expertise did not suffice, what on earth could I tell him about the problems of Eastern Europe?

"You are aware that tiny problems in that region can, notoriously, act as flashpoints for wider European confrontations. Recently such a problem has been giving me grave concern. A tiny island in the Black Sea, Scilla, is currently in disputed ownership between no fewer than three nations: Malbar, Sardina, and Hespia."

"Certainly," I said with renewed confidence. "I have read of it in the newspapers several times recently."

"Capital! Now, in principle the nations are prepared to divide the island into three equal portions, but the question is: what exactly constitutes an equal portion? An arbitrary division into equal areas will not do, for there are various resources to be taken into account. Negotiations have bogged down, and it is causing my department great worry.

"Recently I received a letter from one Reverend Dodgson. He claims that he has an invention that can resolve any dispute of this kind. In fact, he claims to be working on a system that can resolve any dispute over property in a manner acceptable to all parties. He wishes to offer it to the Government free of charge.

"My Department receives many similar letters from cranks, and we normally ignore them. But Dodgson claims to have helped Dr. Watson in a difficult case involving mathematical logic, with successful results. Is this true, Sherlock?"

His brother nodded. "It is true. Watson had more to do with him than I, however."

"He is a mathematician of great ingenuity!" I said emphatically. "I would suggest that his idea is worth hearing."

Mycroft looked at me hard and nodded. "I will trust your judgment, then. He is here at the Foreign Office today. I had thought to delegate a meeting with him to a subordinate. But we will see him ourselves." He pressed an electric bell twice. Shortly afterward Dodgson was shown in; to my astonishment, he was followed by three small, identically dressed girls in the company of a large woman who was obviously their nanny. Dodgson nodded warmly to myself and Sherlock Holmes but addressed the three of us formally: "I thought I would demonstrate my device with the help of children, for it was a children's game that gave me the idea for it. Gentlemen, are you familiar with the procedure that children all over the world use when some coveted item—a piece of cake, say—is to be divided between two of them?"

"I cut, you choose," I said, for the brothers were silent.

"Exactly! One cuts the cake into two portions, the other gets to choose first between them. It is in the interests of the first child to make the cut as perfectly fair as he can manage it, for if there is any difference between the pieces, the second child will take the larger."

I felt considerable disappointment. Had Dodgson come up to London merely to remind us of this ancient method?

"Now the problem," continued the Reverend, "is that the method breaks down when more than two children are involved. Pairs of children the world over must have invented the 'I cut, you choose' method a million times, yet as far as I know, there has never been a way to extend it to even three children."

"How vividly I remember," I said. "I have three nephews of similar age, and when I take them out for the day, the arguments over this kind of thing are—"

Mycroft coughed loudly. "But you think you have found a way, Reverend?" he said.

Dodgson nodded. With great pride, he removed from a large duffel bag an object that was obviously a modified battery gramophone. In place of the playing needle was a kind of guillotine arrangement. He uncoiled three pieces of flex to which were attached pushbuttons and placed the object on the side table, around which the three girls were now sitting.

"When I start the machine, the turntable revolves, but very slowly, about once per minute. Each of the three buttons, when pressed, causes the knife to descend and also illuminates a corresponding bulb so that it is obvious which button was pressed." He demonstrated, and then carefully placed a small cake on the center of the turntable.

"The three young ladies, who have kindly agreed to be called Malbar, Sardina, and Hespia for the purposes of this demonstration"—the three girls giggled demurely—"will represent their countries. The cake could be replaced by a map of the island in dispute."

He manually lowered the knife, making a neat cut to the exact center of the cake, and started the mechanism. The cake turned slowly, each girl watching it intently as she held her button.

"The first girl to press her button will get the slice of cake from the cut I made to the cut she makes. It is in each girl's interest not to press until she thinks the slice is about to exceed one-third of the whole. But after that point she will want to press as soon as possible, lest a rival press quicker and get the increasingly larger-than-fair piece."

Sure enough, when the cake had turned about one-third of the way, the three girls pressed almost simultaneously. Hespia's light came on to show she had been first by a fraction of a second, and she was given the piece of cake the guillotine had cut. Malbar and Sardina watched until the remainder was bisected; this time Malbar pressed first. The three girls paused to curtsy proudly toward us before settling to eat their respective slices.

"Very impressive, Reverend," said Mycroft dryly. "I would like to make one further test before proceeding, though. These well-trained young ladies are obviously delighted to demonstrate the device, but tell me, have you ever tried it with less docile children—boys, for example?"

Dodgson shuddered. "I am well known to be fond of children, except boys," he said. "But I am sure my system will work. The vital point is that it harnesses the players' own self-interest, rather than relying on their altruism, to ensure an equitable division. So it will work equally well with the most unruly boys, and indeed even with warring countries."

At a sign from Mycroft, the girls and their nanny departed, and the three Baker Street urchins we had brought with us were shown in. Dodgson stiffly explained the working of the device to them, and a fresh cake was placed on the turntable and set in motion. The three boys watched, hunched over like racing jockeys. When the cake had turned a third, things happened almost too rapidly to keep track. One boy pressed his button, but his neighbor's hand shot out and pushed the table, dislodging the cake enough to reduce the slice considerably from its expected size. The first boy grabbed it, but in retaliation "accidentally on purpose" knocked the buzzer out of the second boy's hands, breaking its light bulb. The knife had not quite made a clean cut, and the first boy used this as an excuse to scoop away an extra bit at the base.

Moments later, fists and cake were flying. It took the combined efforts of all the adults present to subdue and eject the three lads. The cake had to be confiscated, the glass fragments flung from the bulb having rendered all of it too dangerous to eat. Breathing heavily, Mycroft resumed his seat, combing cake crumbs from his hair. He gazed in dismay at the exquisite light cream carpet.

"Next time you demonstrate the point, brother, I should perhaps suggest a dry sponge cake, rather than chocolate fudge," said Holmes calmly. "Yes, I really think sponge would have been better."

"Lacking your practical experience with these scallywags, I did not anticipate things would get quite so far out of hand," said Mycroft bitterly. He turned to Dodgson. "Nevertheless, you see my point—"

"But they cheated!" protested the Reverend.

"Quite so. The fact of the matter is that humans are disputatious, prone to cheat, and determined to bend any rules that cannot be broken, to gain whatever advantage is to be had. You have only to watch a boys' game such as football to understand that pushing the rules to the limit and testing the referee's discretion is in part what the game is about.

"I fear that a machine that cannot preserve order at a children's tea party is unlikely to contribute much to world peace. I suggest you keep to children's stories and recreational mathematics, Reverend, at which I understand you are quite brilliant."

Dodgson sat crestfallen for a moment, then visibly pulled himself together. "I will take your advice, sir." He gestured at the machine. "Dispose of that as you will. Perhaps someone will rediscover the idea, and improve upon it, in due time."

When the dejected Reverend had left us, Mycroft turned to his brother. "Now to more serious matters. Unfortunately, Dodgson is not the only would-be social reformer in England. Like you, Sherlock, I read the obituary columns, and I have recently noticed some most disturbing patterns. I refer in particular to this series of apparent suicides where harmless sweets are found in front of the victim."

"I have been investigating those," said my colleague.

"Ah! Then the significance of the different numbers of sweets found by each body will not have escaped you?"

My friend looked blank, and Mycroft gave a satisfied chuckle. "You really should get into the habit of tabulating your data, Sherlock. Take all the numeric information available and lay it out clearly, thus."

He passed across a typewritten list. My friend looked at it for a few seconds, then clapped a hand to his head.

"Why, how blindingly obvious! I am quite ashamed. Watson, you should be able to see the pattern at once." He passed the list across.

	Crime	Prison (years)	Age (years)	Sweets Found
Albert Forsyth	Burglary	2	51	9
Derek Carstairs	Robbery	10	29	3
Julia Neilson	Vice	4	45	5
Eli Squires	Forgery	5	34	6

Crime and Punishment

Mycroft rose. "Our next guest is quite distinguished; I think I had better escort him in myself."

I stared at the list, but could make nothing of it at all. Shortly Mycroft returned, accompanied by a tall man dressed in a pinstripe suit with a red carnation. Holmes and I rose to our feet automatically.

"Allow me to introduce the Honorable George White, Director of Prosecutions for the London area," said Mycroft grandly. "Pray be seated, gentlemen. I understand, sir, that you are active in prison reform," he continued, addressing our visitor.

White nodded. "I certainly am. Prior to my present post, I served as Her Majesty's official Inspector of Prisons. They are inhuman and degrading places, where those serving time are treated always with contempt, and sometimes with downright brutality. Even a short sentence in such a place puts a mark on a man from which few can recover. Really, I sometimes think it would be kinder to kill a person than to send him to his first sojourn in such a place. These abysses do nothing to reform the convict and everything to distance him from any hope of an honest role in society."

"You advocate improving prison conditions, then?"

"I used to. But I have come to despair whether this is really achievable. I am now in favor of using alternatives to prison sentences, wherever possible."

This seemed to me a laudable sentiment, but Mycroft frowned.

"A realistic alternative for minor misdemeanors, up to petty thieves and pickpockets perhaps, but hardly for more serious categories of criminals," he said.

"Alas, yes."

Mycroft seemed to consider. "How wonderful it would be, though, if you could invent a way to make prisons obsolete. If you could invent an alternative suitable for every case."

White seemed suddenly very much on his guard. "Well, yes—I suppose it would be," he replied with an unconvincing chuckle.

"It would benefit not only the criminal, but also the public purse. It costs little to hang a man, but a great deal to keep him locked up for years. Why, a man who invented a way to punish crimes intermediate between petty misdeeds and murder that was no drain on our taxes would be lauded as a hero," Mycroft continued.

White barely nodded, staring at him intently.

"Here is a way. Suppose you had caught a man aged fifty. If we take life expectancy as the traditional threescore years and ten, then he has another twenty years to look forward to. Suppose he deserves five years in prison," Mycroft mused, "one-quarter of his remaining life. I have an alternative. Why not force him instead to play a game where he has one chance in four of immediate death—and three chances in four of going free. Perfectly fair odds! Average loss of enjoyable life: five years. But without the hopelessness, the drawn-out suffering, the expense of prison.

"How would you force the gamble? Why, you could sit the man down, show him four candies, one of which was poisoned, and tell him that by picking and eating one, he can discharge his debt to society!"

Mycroft pushed toward White the list he had shown us earlier. "Allowing for the fact that one sweet had been eaten, and rounding off the odds, the number of sweets you offered each victim corresponded exactly to their life expectancy, divided by their expected prison sentence. The pattern is clear."

White slumped in despair. "I suppose one of the others talked. And you deduced I was the perpetrator, despite my disguise. Well, I knew it was a danger, but I was honor-bound to let them go."

"The others?" I said.

Sherlock Holmes smiled. "Of course! We have an observer effect here, Watson. The bodies found correspond to criminals who were unlucky. There must have been many others who were lucky, picked a harmless sweet, and survived. Of course they could not complain to the police about their peculiar ordeal: they would have ended up going to prison after all, and their strange story would probably not be believed. I would guess there must be dozens who escaped unscathed."

White nodded. "Twenty-three, in fact. It would have been more, but I decided to concentrate on more serious crimes, where the expected prison sentence, and hence the probability of death, was high. I suppose someone was bound to blab, in the end."

"No one talked, White. My brother deduced everything by his own powers. But now you have confessed."

White groaned. "Before I am taken away, may I say a few words?"

I was outraged. "You force men to take cyanide, and then expect our indulgence?" I said.

White looked at me, and shook his head. "You are wrong there. I forced nobody. When I was first appointed prosecutor, the case of Annie Footson came up." He gave a short laugh. "Immoral earnings, and she faced a substantial sentence. What hypocrisy—I wonder how many lawyers and judges have dallied under her roof. And prison is even more degrading for women than for men.

"I disguised myself and went to see her. I told her I was in a position to quash the case against her, but my conscience would not let me do so unconditionally. I gave her the choice. She could pick one of ten sweets, one of which was poisoned, and eat the chosen one in my presence. If she survived, that would be the end of the matter. If she refused, the law would take its course in the usual way. She unhesitatingly chose a sweet, and she survived.

"Every subsequent case has been the same. I always gave the criminal a choice. And every one of them, after considering the alternatives, chose to gamble. Does that not tell you something about the merits of my system?

"The first actual death was something of a shock. But by then my mind was made up. I was engaged in a crusade, gentlemen, a moral crusade. The implications if my system is adopted worldwide are quite boundless, more than you might at first suppose. Why, it could soon lead to the end of all property crimes, such as burglary and theft."

I frowned. "I do not quite see—"

"A robber will know that he has a certain chance of being caught and facing my statistical justice. The more he has robbed, the higher his chance of death will be: so much per pound stolen. What I propose to do is to offer a discount! Instead of committing a robbery, he may come to a state-run casino and gamble his life directly for money. Any citizen will be free to do so.

"That also solves the problem of beggars and the poor. It has been suggested that such people should be paid a basic living from the public purse. I regard that as a mere encouragement to idleness. When indigents complain of having no money, why, they are free to go to the public casino and gamble for as much as they want! There will be neither excuse nor incentive for beggary or theft, for the casino will offer more favorable odds than the punishment for those crimes."

White flung his arms wide. Flecks of foam showed at the corners of his mouth.

"I see a vision of the brave new world outlined by Wells and others. But there are no prisons in this world, and no poor-houses. Instead, the masses gather before a simple glass booth in the town square. Before the booth stands a queue: criminals, the indigent, those so lazy they prefer gambling to hard work—in short, all those who are a burden to society.

"One at a time, they step into the booth and grasp two copper handles. A great roulette wheel spins, set to appropriate odds in each case. Most of the time, only a small current flows, just enough to verify that the accused has grasped the handles firmly. A green light shines: the accused may go free. But every now and again, there is instead a great red flash: a huge voltage direct from the power station has reduced the wretch to a heap of ashes, instantly and painlessly. What a wonderfully clean and pure world this will be! And it will get better and better as time goes by, for with the undesirables of every generation systematically pruned, eugenic improvement will inevitably occur."

Mycroft shook his head. "Your motives may conceivably be good in your own eyes," he said ponderously. "But that does not excuse you. Taking the law into your own hands, as you have done, is a crime both in law and against natural justice."

White smiled. "Remember, I am a lawyer," he said. "I have given some thought to what crime you could charge me with. Not murder—my clients all took their chance voluntarily. Not blackmail—I have not gained by my actions. I doubt you can even make an assisted-suicide charge stick, for each client stood a better-than-even chance of survival."

"Conspiracy to pervert the course of justice," said Mycroft simply.

White sighed. "Yes, there is that."

"I fear that as a former judge, you will be among those who do not have an easy time of it in prison."

White drew his handkerchief and used it to wipe the foam from his lips. "I shall not go to prison. Every great cause needs a martyr," he said indistinctly.

My colleague sprang up, but White shook his head. "Too late!" he said. "I foresaw this moment, and I always carry one of my sweets in my pocket. Just remember this: every one of my clients took his medicine voluntarily. *Even the criminals perceived the greater merits of my system.*"

Then he started to cough and choke horribly. In a few moments it was over.

We repaired to a less grand room while the body was attended to; a respectful police superintendent came to take brief statements from us. It seemed to me that the time had come for us to leave, and I stood up. Mycroft waved me back to my chair.

"Sit down, Doctor. I am afraid we are not finished yet. Not all the suspicious deaths I have noticed recently have been White's work. There have been others. We have met a priest and a lawyer this afternoon, but who is better placed to commit the perfect murder than a member of the third of the traditional professions: a medical doctor?"

His finger stabbed out at me. "I have read the obituaries of too many of your patients recently, Doctor. Yesterday I prepared to have a warrant issued for your arrest."

I felt icy fingers on my spine. Sherlock Holmes was not above the occasional practical joke, but from his brother a jest was inconceivable. Indeed, Mycroft's stony face showed his seriousness.

"Then I realized I was overlooking another possibility. Parker, show in our next guest."

A uniformed attendant hastened away. Moments later, and to my utter astonishment, he returned with my partner, Penbury. There must be some mistake here: Penbury was perhaps not above taking advantage of me, but I knew he was a conscientious doctor. Had the Holmes brothers decided to frighten him for my benefit, as some quite excessive revenge for the matter of the bets on the horse race? I rose to protest, but Mycroft spoke first.

"Dr. Penbury, I believe these tablets bear your surgery stamp, showing that you made them up personally." He held out a stoppered glass bottle of small blue pills that I knew only too well. "Such were found by the bedside of all those who have recently died of Baird's disease in London. I have had the police laboratory analyze them. They are supposed to contain an extract of nightshade. The laboratory found that they were only chalk and water. Would you care to explain?"

To my horror, my partner made no attempt to deny the charge but merely nodded. He took a seat uninvited, his head at an arrogant angle. "Certainly, gentlemen. Have you heard the old joke, that every doctor knows that half the medicines he prescribes do no good, and indeed may harm his patients? The trouble is, he does not know *which* half! I am afraid there is much truth in it. One's subjective impressions can be quite deceptive.

"Recently John here believed he had noticed that nightshade extract cured Baird's disease. But was this true or merely wishful thinking? After all, cases of Baird's quite often recover spontaneously. We really needed to know. After all, nightshade extract is potentially poisonous; you would not want to prescribe it unnecessarily. I thought of a perfect way to find out—a way that would eliminate subjective judgment.

"I divided our patients randomly into two groups. I gave Watson identical-looking pills for both groups. Unknown to him, the first group received real medicine, but the second only chalk."

He looked at me. "It was you who gave me the idea, Watson, when you explained how you could tell whether a coin was biased only by tossing it many times. I have given one hundred patients the real medicine and one hundred the harmless placebos. In a year's time, we will see how many of each group survive and for the first time in history will have unequivocal statistical proof that a medicine really works!

"I dub it the double-blind test. Neither doctor nor patient knows who is getting real medicine and who is not, so placebo effects and wishful thinking—sometimes a patient can get better or worse just because of his own expectations, or those of his doctor—are automatically eliminated."

I could contain myself no longer. "Penbury, what you have done would be wrong in any case. But after the first few times, your actions became totally unnecessary. Every single patient who got the real drug survived; all the others died. Why did you not abandon this madness when that became clear?"

Penbury jutted out his jaw. "I could not be sure of the odds. I knew how to calculate the statistical significance of the test once it was complete. But obviously, during the trials, the results might fluctuate up and down. I did not know how to calculate the chance that the results might appear spuriously good at some intermediate point. For full accuracy, it was necessary to run the thing to the end, even if weaker-willed men"—here he looked scornfully at me—"would have faltered in their resolve."

"Did it not occur to you to consult a statistician, using some cover story to prevent his reporting you?" asked Mycroft softly. "Those were human lives, not poker chips, that you were playing with."

Penbury drew himself up grandly. "I am a member of the medical profession, and we defer to nobody!" he said. "The ends justify the means. Where would modern medicine be without vaccination? We have that only because a century ago Jenner injected a healthy child with smallpox, on the off chance that a previous cowpox infection would protect him, while the child's mother looked on trustingly. Our patients are ours to use for the greater good, even as a general expends his men in battle. I am under no obligation to explain my actions to mere laymen!"

He would say nothing further. Eventually Mycroft shook his head sadly and pressed an electric bell three times. My erstwhile partner was led away between two policemen, to face a charge of multiple murder.

After the cool of the Foreign Office interior, the late afternoon air was positively stifling. We walked at a somber pace toward the corner of Hyde Park.

"An eccentric priest, a mad lawyer, and a madder doctor," I said eventually. "The professions that laymen put their trust in have not exactly been shown up to advantage. My own came out worst of all. The priest did not set out to harm anybody, and the lawyer only those who had committed serious crimes. But Penbury sentenced innocents to die, as arbitrarily as by tossing a coin. At least I am quite sure his system will never be adopted by my colleagues."

Mycroft looked at me. "I would not be so certain," he said. "It is indeed necessary to validate new medicines. And the public places such trust in doctors that they can literally get away with murder. I am not confident that the charges against Penbury will stick. I think the best we can hope for is that your colleagues will not be too arrogant to ask the mathematicians for help. If they are left to administer such systems by themselves, they may expend far more lives than necessary— like incompetent generals who do not plan their strategy, but instead sacrifice foot soldiers by the thousands, in the hope of overwhelming the enemy by sheer numbers."

I was appalled at this comparison, and for a while we walked in silence. At the corner of the Park, Sherlock Holmes paused. "The side road is more direct, but the route across the grass much more pleasant," he said. "I would vote for the latter. How say you, Mycroft?"

Mycroft looked doubtfully up at the sky, which in the last little while had changed from clear blue to a threatening bronze.

"I see a thunderstorm coming. I would prefer the direct route," he said.

"Well, it seems we have a tie," said my friend cheerfully. "It is up to you to break it, Watson."

I hesitated and looked about. As I spoke, the heavy atmosphere seemed to reflect my words back at me, so that the lightest comment seemed imbued with a massive significance.

"I think the storm is some way off yet, and when it comes, everyone will be cowering indoors for quite some time; it looks to be a big one," I said. "Let us gather our rosebuds while we may, and take the Park route."

Mycroft assented reluctantly. As we walked into the Park, we came within sight of Speaker's Corner, where deranged but sincere individuals stand on soap boxes to preach their ideas to a generally jeering crowd.

"I suppose they are always at least half-mad, these visionary reformers," I said sadly. "Probably any individual who thinks he can change the ways of the world is deluded."

"Not at all," Mycroft said firmly. "The modern world is utterly different from the brutish one that our ancestors once inhabited. Most of the improvements required a visionary to suggest them. And some of those visionaries are rightly remembered as the greatest names in history."

"But surely you would not support this statistical justice idea?"

Mycroft shook his great head. "No. Doctor, you have come to understand the laws of probability, which apply when you are playing against the laws of the Universe. You have also met its more subtle refinement, game theory, which applies when you are playing against an opponent who can change his strategy to outwit you. To predict the result of a social innovation requires a kind of meta-game theory: you must predict its impact on a society made up of individuals and groups who will themselves be formulating plans to exploit the new situation.

"Statistical justice might be workable if enforced by angels. But there are too many temptations. White himself described how, as well as punishing crimes, his machines might be used deliberately to lure the needy and easily tempted into oblivion. When it is so cheap and easy to punish, the definition of crime can too easily be extended. Who comes next, unpopular groups like the travelers, the Gypsies?

"At the moment there is at least one factor that deters governments from punishing too many people: the high cost of imprisonment. Even the greatest dictator in history has been constrained by the cost of running his prisons; it is impractical to jail more than a tiny fraction of the whole population. No, I foresee many horrors in this new century, but I do not think that particular idea will come to pass."

We paused to buy chilled cups of lime juice at the stand in the center of the park, and sipped our refreshment as we walked.

"I fear there is a long history of well-meaning but impractical persons proposing improvements to society without considering the real-life problems," continued Mycroft. "My favorite example is the late Marquis de Condorcet. Intrigued by the work of French mathematicians such as Pascal, he suggested that criminal trials should always use a large jury to reduce the chance of bias. Then you could have confidence in the verdict.

"Come the revolution, he was himself placed on trial on trumped-up charges, before a large panel of judges, all with equally extreme views, who unanimously found him guilty. He was guillotined. A wonderful example of poetic justice, to my mind. A warning to theorists who would dabble in changing society's rules."

I shook my head in despair. "You think it is hopeless to use scientific methods such as game theory to devise improvements to society, then?" I said.

"On the contrary. Conforming to scientific logic is *necessary*, but it is not *sufficient*," said Mycroft. "Take the Reverend's idea. His intuition that you will get a fair division of goods only if you harness the interests of the individual was perfectly correct. He was certainly more realistic than socialist visionaries such as Wells, who seem to believe that in a perfect society everyone will work hard and share goods fairly out of the goodness of their hearts. Unfortunately, his cake-cutting machine was not a sufficiently foolproof solution to the problem of fair division.

"In fact, I have the utmost sympathy with those trying to devise a better society. Laissez-faire is not the answer. If one set of rules might lead to a society where everyone is idle, another could lead to everyone working with ever more desperate futility, like a caged hamster running on its wheel. I anticipate that in a hundred years' time, machines will be so effective that the necessities of life could be earned by just a few minutes' work each week."

"At last everyone will have time for study, for meditation, for philosophical discourse," I mused.

"I doubt it, Doctor. If we are not clever enough to frame the right rules, we could even end up working harder than today, women as well as men clamoring to labor every available hour. But they will work for firms striving harder to outwit one another than to produce useful goods. Almost all the work will be unnecessary, doing complicated work to make complications for others, a self-perpetuating headache.

"We may well fall into that trap. But my still greater fear is that by proposing 'scientific' rules for society without considering the full consequences in human terms—that is, without taking account of game theory and also of meta-game theo-

ry—true horrors may result. Economic reform, crime and punishment, medicine, eugenics—all the well-meaning proposals
conceal possibilities for abuse. The twentieth century could
end up being remembered as much for its human failures as
for its technological progress. The whole concept of radical
improvement of human institutions could become so discred-
ited that it will no longer be permissible even to discuss such
ideas in polite society."

The idea that the world might not have improved in a century's time struck me as so ludicrous that I did not bother to
reply. It seemed to me that the marvelous advances of science
would improve the human condition automatically, the rules
of society being mere details.

Mrs. Hudson met us at the door in something of a confusion. "Oh, sirs, you have missed them," she exclaimed. "By
just a few minutes. And them so anxious to talk to you—the
older one, especially."

"To whom are you referring?" asked Sherlock Holmes cour-
teously.

"I did not ask their names, but they did leave their cards. A
very old man, and a very young one. I showed them into the
parlor, and told them you would not be long." She blushed
slightly. "In fact, I may have overheard a little bit of their conversation; I happened to be dusting in the hallway, you see.
The old man was saying that he wished these new branches of
mathematics—I think he spoke of something called game theory, and certainly a lot about economics, though the details
went over my head—he wished they had been invented when
he was writing his book. There were various things he now
thought it necessary to take into account. He was very eager
for your advice, Mr. Mycroft, sir.

"The young man listened with great respect. But the impa-
tience of youth was upon him. He kept pacing up and down,
saying there was a train he had to catch—apparently he had to
go to attend some meeting in Paris." She lowered her voice to

a whisper. "In fact I distinctly heard him say something about a *revolutionary* meeting, begging your pardon. In the end they both left just before you got back."

"Do you have the cards, Mrs. Hudson?" asked Mycroft patiently. He took them from the proffered salver, and raised his eyebrows.

"Dear me! It would seem that none other than the famous economist Karl Marx desired to consult us."

"What is the other name?" asked Holmes.

Mycroft shook his head. "It means nothing to me. Unusual— my Department makes a point of keeping an eye on these young hotheads. Perhaps it is a pseudonym.

"His card describes him as Lenin. Vladimir Ilyich Lenin."

Afterword

I HAVE TRIED TO KEEP the main part of this book at a level relevant to everyday life, sampling many areas of math, logic, and decision theory without getting too formal, or going too deeply into any one topic. In case I have piqued your curiosity, however, here are some additional comments and suggestions for further reading. For up-to-date information, see my website: http://members.aol.com/OxMathDes/ColinBruce.html.

First, a note on historical accuracy. The contents of the book are of course pure fiction, even though some historical characters appear. The book is set in 1900; the Reverend Charles Dodgson (alias Lewis Carroll) really died in 1898. He wrote the Alice books, and devised mathematical puzzles including the one of the black and white balls, decades earlier than implied here. Marx died in 1883, when Lenin was only 13, so they could never have met as adults, as described. I hope the reader will forgive my taking these liberties.

Chapter 1, *The Case of the Unfortunate Businessman*, focuses on three well-known business errors: the cab driver's fallacy, the prior investment fallacy, and the fallacy of mistaking relative for absolute savings. I chose these particular examples because they are as relevant to the humblest household decision as to the largest corporate one, and because each can result from the *mis*application of a perfectly sound business principle. The essential practice of setting goals for yourself or other people can result in the cab driver's error: wasting effort in some predetermined direction, while ignoring other opportunities. The desire to make everything you do succeed can cause the prior investment error: throwing good money or effort after bad in an attempt to recoup. The desire to economize can cause the relative/absolute savings error (traditionally known as penny-wise, pound-foolish): using disproportionate effort to save trivial amounts of money. For a more comprehensive discussion of business decision errors, I recommend Russo and Schoemaker's book *Decision Traps*.[1]

On a lighter note, the criminal scam that kicks off Chapter 1 is based on David Maurer's 1940 classic, *The Big Con*.[2] The kind of grand-scale con game described in this supposedly factual book has inspired much fiction, most famously the film *The Sting*. *The Big Con* is still in print and still wonderful, but be warned that it almost certainly also demonstrates a basic point about statistics: you must base them on reliable data. Maurer's research method was essentially to interview retired con men, applauding them in proportion to the outrageousness of their stories. They all seemed to be living in strangely reduced circumstances for such successful criminals. The accounts must be taken with a large pinch of salt, but some aspects do match the stories of friends of friends of mine who have been conned in the present day. A key point is that the opportunity offered you will involve something that initially appears only mildly illegal, not enough to frighten you off, but then comes to seem more serious, so that you are reluctant to go to the police when things turn sour. It will appear that everybody lost, not just you, so although you may have your suspicions, it will be more comfortable to assume you were a partner in a scheme that went wrong than to face the fact that you alone have been had.

Chapter 2, *The Case of the Gambling Nobleman*, describes the two most common gambling fallacies: the notion that the laws of probability make a coin more likely to come up heads after a run of tails, and the idea that you can ignore an infinitesimally small chance of an infinitely large loss, or indeed a very small but finite chance of a very large loss. For a detailed description of strategies for the most widely available forms of commercial gambling you could read John Haigh's *Taking Chances*,[3] but the best advice is still summarized in one word: don't!

Chapter 3, *The Case of the Surprise Heir*, is about more than the well-known birthday paradox. The important points are that coincidences can be much less unlikely than they appear, and that genuine correlations often seem to imply that a spurious cause-and-effect relationship exists. These are major factors in reinforcing superstitious beliefs.

Chapter 4, *The Case of the Ancient Mariner*, introduces the concept of probability distributions, including the most famous example: the Normal Distribution, or bell curve. The triangle of numbers on page 75 is properly called Pascal's Triangle, and has many applications in applied mathematics. Any textbook of probability and statistics will cover these topics, but not all point out the mathematically beautiful connection between the Drunkard's Walk, Pascal's Triangle, and the Normal Distribution, which to my mind is the best way to see why so many populations of objects both man-made and natural match the bell curve so closely. It is a subtle property of our universe that there tend to be more ways for multiple influences to cancel than to reinforce one another: this is responsible for the relative uniformity and predictability of the world about us.

The mistake Watson makes at the end of this chapter, when trying to decide what probability threshold it is appropriate to use in a given context, is a common one. A clever manufacturer's spokesman will insinuate that no one should stop using his product until it is *proved* to be dangerous, implying that a very high level of evidence is required. This is absolute nonsense. A chemical additive, for example, does not have any right to the presumption of innocence that we would grant a human defendant! You are quite entitled to take very small risks into account in your everyday decisions, especially if you can avoid the danger at little cost or inconvenience to yourself.

The common factor of the several puzzles in Chapter 5, *The Case of the Unmarked Graves*, is that almost everybody finds them much harder than they really are. The first two puzzles, of the stones marked with circles and lines, and the youths who must not be allowed alcohol if they are under age, together constitute a *Wason test*. Wason's hypothesis is that for evolutionary reasons, human brains are hard-wired to detect cheating. Hence we find the cheat-detection problem much easier than an equivalent abstract problem. In a standard Wason test, the cheat-detection part of the problem is as I have described it, but the abstract version must be disguised so that subjects do not catch on that it is really the same puzzle. It is therefore presented as a set of cards with letters on one side and numbers on the other. The subject must turn over cards to test a proposition such as "Any card with an odd number on one side has a vowel on the other." To my mind, this difference may introduce other difficulty factors. I have used the flim-flam of narrative to distract the reader so that the "abstract" and "cheat-detection" versions of my test are more nearly identical, without (I hope) giving the game away. Was the abstract version really so much harder? What do you think? You can read about more conventional Wason tests in Stephen Pinker's *How the Mind Works*,[4] Matt Ridley's *The Origins of Virtue*,[5] or Brian Butterworth's *What Counts*.[6]

The many-worlds tree shown on page 105 is more properly called a probability tree. I find it useful to invoke the notion from quantum physics that we may live in a superposition of realities, in which every physically possible outcome actually happens and is equally "real." Doing so emphasizes that to assess posterior probabilities correctly, we must consider whole branches of might-have-beens, looking in detail at their cascading development, even when we know the event that would have made them possible did not actually occur in "our" world. Of course, the usefulness of this mental technique does not depend on the correctness or otherwise of this controversial physical theory. For a description of this and other puzzles of modern physics in a format similar to the book you have just read, see my previous book *The Einstein Paradox, And Other Science Mysteries Solved by Sherlock Holmes.*[7]

In Chapter 6, *The Case of the Martian Invasion*, I have caricatured the long tradition of finding coded messages or prophesies hidden in religious texts. The point is, of course, that the number of permutations available in any large text is so vast that it is possible to find almost any message you want, especially now that computers are available to help in the search. Incidentally, using the Hebrew version of the Bible as your source material makes it much easier to find proper names. When you transcribe names from one alphabet into another, there is no uniquely correct way to do it. An example is the Russian mathematician who helped lay the foundations of statistics. His name is a nightmare to look up in English indexes: it may be translated equally correctly as Chebyshev, Shebychev, Tschebishev. . . . Moreover, in ancient Hebrew as a general rule only the consonants are written; vowels must be inserted by the reader. To find the name "Sherlock Holmes" correctly spelled in the English version of the Bible would indeed be startling, but to find a string of consonant sounds that, with any vowels you wish inserted, is a plausible phonetic rendering is of course vastly easier.

The weed-killer explanation for the origin of the crop circle phenomenon is my own. It occurred to me when a friend told me of the trouble he had keeping his garden orchard clear of pests, despite using all the legally permissible sprays he could find. He was puzzled to see that the same trees on adjacent commercial orchards were pest-free. When he went to visit his farmer neighbors, there were always drums of chemicals lying around, strong pesticides that it was illegal to use on soft-skinned fruit—or indeed any crop his neighbors grew. My idea is no more than a guess, but I find it more plausible than hypotheses put forward by well-meaning physicists involving plasma vortices or microtornadoes. If such phenomena occurred regularly, it would be hard to imagine their escaping notice in the crowded land of England! Scientists can be as gullible as anybody else, and they sometimes forget to take into account that people have many possible motives for lying. In fact, professional conjurers—individuals such as James Randi and groups such as the British Magic Circle—have been far more successful than scientists at exposing bogus "phenomena."

The rather alarming statistics on aircraft engine failure are of course fictional, although they might have been reasonable for early internal combustion engines. Modern jet engines have a failure rate per flight nearer 1 in 100,000 than 1 in 100. But the basic point about overrating safety is a real one. It is incredibly unlikely that any four-engine aircraft will lose all four engines for different and unrelated reasons, but I know of at least two cases where modern four-engine aircraft have suffered total power failure. One was a Boeing 747 that flew through a cloud of volcanic dust, causing all its engines to flame out. The other was a plane that took off on a test flight without passengers after major maintenance. An engineer had left oil drainage valves on all four engines in the wrong position, and they started to fail in rapid succession. Incidentally, both planes made emergency landings without loss of life. Such multiple engine failure is very rare, but not as unlikely as it would be if the probabilities were truly independent. Similarly, although nuclear power stations are safe by ordinary standards, multiple backups do not multiply the chance of meltdown into insignificance if all the safety systems share the same flaw, which has gone undetected because they have never been properly tested; or if someone turns all but one of the systems off to perform a thorough safety test, as happened at Chernobyl.

Chapter 7, *Three Cases of Unfair Preferment*, describes the error of assuming that a well-defined ordering relation must also define a unique hierarchy. In the world of ordinary numbers, we know that if x is greater than y and y is greater than z, then x is greater than z. But in higher mathematics it is quite possible to have x greater than y, y greater than z, and yet z greater than x! Some psychologists claim that our ingrained unique-hierarchy assumption comes from a notion of tribal pecking order: if A can thump B, and B can thump me, then I had better not get in a fight with A. As far as I know, there is no real evidence for this, and it seems to me more likely that we are programmed to assume hierarchy simply because it is very often true when we compare such properties as size, weight, temperature, and so on: there is usually a simple and unique ordering system for any ordinary set of objects, such as food items. Baked Alaska—one item of food at two different temperatures—is the exception that proves the rule.

Chapter 8, *The Execution of Andrews*, introduces Bayesian logic. In certain circumstances, intuition leads us to use entirely inappropriate ratios for calculating probabilities. An extreme case would be a police interrogator reasoning as follows: "You deny committing the crime. Ninety-nine percent of criminals deny their guilt. Therefore, it is 99 percent likely that you are guilty." The examples given are of practical importance. Surveys show that something like 95 percent of doctors make the same error as Watson, grossly overestimating the chance that a patient with a positive test result actually has a disease that is known to be rare. Juries in criminal trials are probably even more easily misled. Incredibly, in England lawyers are currently forbidden to explain Bayesian statistics to jurors, on the basis that it is likely to confuse them! As I write, however, a case is going to appeal in England that perfectly illustrates the importance both of Bayesian reasoning and of the fallacy of wrongly assuming that probabilities are independent. A woman was found guilty of murdering her two children on the basis that the chance of their both having suffered genuine sudden infant death syndrome (or SIDS, commonly called crib death) was "73 million to 1 against." Presumably this figure was arrived at by multiplying 1/8,500 by 1/8,500, this being the chance that an individual child in a family with no particular risk factors (such as poverty or heavy smoking) dies of crib death. First, whatever the genetic or environmental factors behind crib death, two children of the same family inevitably share many of them. And second, Bayesian reasoning tells us that even if this huge figure were correct, it is *not* the same as the probability that the accused is guilty.

The Reverend William Bayes lived and died in the eighteenth century, but his insights into probability and decision theory were so profound that conferences on his philosophy, now dubbed Bayesianism, are still held today. He gave us Bayes's theorem, the Bayesian decision rule, the Bayesian decision tree, and the Bayesian probability tree. Some of the more subtle questions he asked remain unanswered and are still keenly debated by mathematicians. Surprisingly, I know of no readily available popular book about this remarkable man and his philosophy, although *Bayesian Decision Problems and Makov Chains,* by J. J. Martin,[8] is a good starting textbook.

Chapter 9, *Three Cases of Relative Honor*, introduces the tricky field of game theory. In reality, it was not until the middle of the twentieth century that this discipline got properly under way, pioneered by mathematicians such as von Neumann, who also famously contributed to computer science. Intriguingly, one of the first situations von Neumann chose to model was Sherlock Holmes's Canterbury-or-Dover dilemma in *The Final Problem!* (For the benefit of diehard readers who look up von Neumann and Morgenstern's original paper: I get a slightly different result, not because either of us is in error, but because I assigned different values to the desirability of the possible outcomes to make the arithmetic easier to follow.)

Game theorists initially found it very difficult to understand why people so often cooperate and behave honestly, even in Prisoner's Dilemma-type situations. The breakthrough came with the realization that real life consists not of single set-piece "games," but of indefinite iterations. Those who employ selfish tactics tend to gain in the short term but to lose in the long run. Fortunately, the early findings of game theory were not taken too seriously. It is rumored that during the Cold War years, Kennedy and his successors intuitively rejected suggestions by mathematical strategists that nuclear retaliation should be on a proportionate basis: you nuke one of our cities, and we'll take out one of yours. Instead he told the Russians that even a single attack on a U.S. city would trigger an all-out response. The consensus is now that an unpredictable, potentially disproportionate response is indeed the best deterrent strategy. This may explain why anger is hard-wired into our emotions in the way it is. The current (revised) edition of Richard Dawkins's 1976 classic *The Selfish Gene*[9] contains a new thought-provoking chapter on game theory and its implications for human and animal behavior. For a fuller account of game theory, see *Game Theory: A Non-Technical Introduction* by Morton Davis.[10]

Chapter 10, *The Case of the Poor Observer*, and Chapter 11, *The Case of the Perfect Accountant*, should be taken together. They deal with the same problem: How do you construct an accurate picture of the world, given that your subjective impressions may be misleading, and second-hand reports deliberately selective? For an explanation of the ways statistics can be misused, Darrell Huff's classic *How to Lie with Statistics*,[11] written half a century ago, is still worth reading. However, it is one of those books (*The Selfish Gene* is another) whose title is often quoted by people who have never read it and are quite mistaken about its real message. You should indeed be wary of information from a biased source: politicians and salespeople know how to distort the truth in any language, including that of mathematics. But audited statistics give a much more reliable picture than, for example, selective anecdotes. On those rare occasions when they actually print statistics, newspapers often get them right. Unfortunately, most people base their picture of reality not on the known statistics, but on how often they read a report of a particular event. Extreme stories sell papers; mundane ones do not. Accordingly, our perception of rare but horrific dangers is greatly exaggerated.

For example, children are tragically accident-prone, and so a story about a child run over by a car is not "news," whereas child murders are always widely reported. I know many intelligent and educated parents who will not let their children walk to school, for fear that they will be abducted and murdered by a maniac, yet allow them to bicycle in traffic. The chance of the child suffering a lethal road accident is *vastly* higher (in Britain about 30 times greater) than that of being murdered by a stranger. Similarly, in Europe intercity buses are many times safer than private cars, because both the vehicles and their drivers are subject to more stringent checks. But car accidents are too common to make the news, whereas serious bus accidents do. As a result, intercity buses have recently been restricted to a top speed about 10 miles per hour lower than cars, making them a less popular travel choice. To minimize road deaths, it would have made more sense to *increase* the speed limit for buses relative to cars, luring more drivers off the road.

I had a recent reminder of how powerful the impact of selective reporting can be when I turned on the television for the evening news headlines. The lead shots all happened to feature major conflagrations: a forest fire, an oil refinery accident, a war zone. Of course, my conscious mind knew there was no connection, but my visceral reaction was "Help! The world's on fire!" I don't see any real cure for the problem. Editors are quite right to suspect that repeated headlines like "No Reported Crime in Your Area Last Night" and "No Serial Killers in Town, but Lots of Bad Drivers: Read This Boring Road Safety Drill, Again" would do little for their circulation figures. We must just do our best to remember that the media relentlessly bombard us not with the typical, but with the most extreme and unusual events from an entire continent.

The law that in a set of data of widely differing size (that is, spanning an order of magnitude or more) numbers are more likely to start with low digits than high ones has been discovered twice: once by Simon Newcomb in 1881, and again by Benford in 1938; it is now known as Benford's Law. It happens because the quantities are more likely to be distributed uniformly when written on a logarithmic rather than a linear scale, as on a slide rule. To put it another way, they are likely to be separated by a typical *ratio* rather than a typical *difference*. To my physicist's mind this is an unsurprising finding, but there is still lively argument between mathematicians as to the precise reason for it. You can find many papers on the Web disputing the theology of the matter.

Of the alarming ideas introduced in Chapter 12, *Three Cases of Good Intentions*, double-blind medical trials have of course become reality. Until quite recently, such trials did indeed often suffer the ethical defects highlighted: large numbers of patients with life-threatening conditions were given placebos and died. I am happy to say things have improved somewhat. Under international guidelines, when a new drug is being tested and an older but effective treatment for the condition is known, the members of the control group are supposed to receive the older drug, rather than a placebo. This is not only ethically more satisfactory: it also markedly reduces the incentive for patients or their doctors to cheat; and because real drugs have side effects, whereas placebos do not, it becomes harder for anyone to tell which group a particular patient is in, reducing the possibility of observer bias. Another change is that if early results show unequivocally that the new drug is better (or worse) than the old, the trial is not continued merely in the interest of generating more precise statistics. Rather, the trial is abandoned, and everyone is given the more effective treatment. But there is still room for improvement: the guidelines are not always followed, and in any case we need algorithms that provably and precisely minimize the number of patients denied the most effective treatment, relative to the true value of the information garnered. Fortunately, mathematicians are now fully involved: I see that at least one new book on the Bayesian analysis of clinical trials is in press as I write.

I have come across variants of the statistical-justice idea several times in science fiction stories, although the scheme presented here is my own. Of course, I am not seriously proposing it. Quite apart from the snags identified by Mycroft, there is the problem that it is impossible to reprieve people who have been executed if new information comes to light. Also, there is plenty of evidence that a small probability of a severe punishment is a less effective psychological deterrent than a high chance of a small penalty. Drunk drivers who are undeterred by a real risk of their own death will give up the habit rather than pay fines of a few dollars. But I wanted to introduce the idea that new solutions to society's problems may be possible.

A more promising approach to this is through game theory. Almost everyone is selfish to a degree, putting self and family ahead of the rest of the community. But if we invent the right rules, this pursuit of self-interest can actually create greater fairness. This is what Mycroft was trying to do in the previous chapter, using a carefully set up game of Prisoner's Dilemma to extract confessions. It is also the idea behind the cake-cutting machine (properly called the moving-finger or moving-knife system): each player pursuing his or her own selfish interest actually results in the fairest possible division. Much more sophisticated fair-division algorithms have now been invented. For an up-to-date account, see *The Win/Win Solution* by Brams and Taylor.[12] There have even been ambitious suggestions that such schemes could bring about progressive world disarmament, each nation giving up only what it regarded as a "fair" proportion of its weapons at each stage. Although Brams and Taylor identify limited successes, for example in the international division of seabed mining rights, more ambitious goals remain elusive. But game theory and related branches of mathematics have made great strides in recent decades. Perhaps where the visionaries of the early twentieth century fell short in their attempts to design new and better societies in which war and want would be unknown, those of the twenty-first, equipped with better knowledge, may yet succeed.

Notes

1. J. Edward Russo and Paul J. H. Schoemaker, *Decision Traps: Ten Barriers to Brilliant Decision-Making and How to Overcome Them* (Doubleday Books, 1989).

2. David W. Maurer, *The Big Con: The Story of the Confidence Man* (Anchor Books, 1940; reprinted 1999).

3. John Haigh, *Taking Chances: Winning with Probability* (Oxford University Press, 2000).

4. Steven Pinker, *How the Mind Works* (W. W. Norton and Co., 1999).

5. Matt Ridley, *The Origins of Virtue: Human Instincts and the Evolution of Co-operation* (Viking Books, 1997).

6. Brian Butterworth, *What Counts: How Every Brain Is Hardwired for Math* (Free Press, 1999).

7. Colin Bruce, *The Einstein Paradox, And Other Science Mysteries Solved by Sherlock Holmes* (Perseus Books, 1998).

8. J. J. Martin, *Bayesian Decision Problems and Makov Chains* (Krieger Publishing Co., 1975).

9. Richard Dawkins, *The Selfish Gene,* 2nd ed. (Oxford University Press, 1989).

10. Morton Davis, *Game Theory: A Nontechnical Introduction* (Dover Publications, 1997).

11. Darrell Huff, *How to Lie with Statistics* (W. W. Norton and Co., 1954; reissued 1993).

12. Steven J. Brams and Alan D. Taylor, *The Win/Win Solution: Guaranteeing Fair Shares to Everybody* (W. W. Norton and Co., 1999).

Index